DAM SURVEILLANCE – LESSONS LEARNT FROM CASE HISTORIES

SURVEILLANCE DES BARRAGES – LEÇONS TIRÉES D'ÉTUDES DE CAS

T0332613

INTERNATIONAL COMMISSION ON LARGE DAMS
COMMISSION INTERNATIONALE DES GRANDS BARRAGES
61, avenue Kléber, 75116 Paris
Téléphone : (33-1) 47 04 17 80
http://www.icold-cigb.org./

Cover/*Couverture* :
Cover illustration: *Zoeknog Dam (South Africa), January 1993 / Barrage de Zoeknog (Afrique du Sud), janvier 1993*

CRC Press/Balkema is an imprint of the Taylor & Francis Group, an informa business
© 2022 ICOLD/CIGB, Paris, France

Typeset by CodeMantra
Published by: CRC Press/Balkema
Schipholweg 107C, 2316 XC Leiden, The Netherlands
e-mail: enquiries@taylorandfrancis.com
www.routledge.com – www.taylorandfrancis.com

Original text in English
French translation by M. Balissat (chap. 1 to 3)
T. Adeline, C. Casteigts, T. Guilloteau, F. Laugier, J.C. Palacios,
G. Pavaday, M. Poupart, G. Prevot & E. Vuillermet (chap. 4 to 6)
Layout by Nathalie Schauner

Texte original en anglais
Traduction en français par - M. Balissat (chap. 1 à 3)
T. Adeline, C. Casteigts, T. Guilloteau, F. Laugier, J.C. Palacios,
G. Pavaday, M. Poupart, G. Prevot et E. Vuillermet (chap. 4 à 6)
Mise en page par Nathalie Schauner

ISBN: 978-1-032–22940-9 (Pbk)
ISBN: 978-1-003–27484-1 (eBook)

LIST OF MEMBERS / LISTE DES MEMBRES

CO-CHAIRMEN / CO-PRÉSIDENTS

South Africa / Afrique du Sud	Chris OOSTHUIZEN († 2017)
Spain / Espagne	Jürgen FLEITZ

MEMBERS / MEMBRES

Argentina / Argentine	Alejandro PUJOL
Australia / Australie	Andrew REYNOLDS (since 2017)
Austria / Autriche	Florian LANDSTORFER
Canada	Pierre CHOQUET
Egypt / Egypte	Ashraf EL-ASHAAL
Czech Republic / République Tchèque	Pavel KRIVKA
France	Thierry GUILLOTEAU
Germany / Allemagne	Markus AUFLEGER
Iran	Ali NOORZAD
Italy / Italie	Alberto MASERA
Japan / Japon	Jun TAKANO
Morocco / Maroc	Ahmed CHRAIBI
Mozambique	Ilídio MARCOS TEMBE
Norway / Norvège	Goranka GRZANIC
Portugal	Carlos PINA
Romania / Roumanie	Iulian Dan ASMAN
Russia / Russie	Viacheslav SOBOLEV
Sweden / Suède	Sam JOHANSSON
Switzerland / Suisse	Laurent MOUVET
USA / États-Unis	Jay STATELER
UK / Royaume-Uni	Ian HOPE

CO-OPTED MEMBERS / MEMBRES CO-OPTES

Colombia / Colombie	Alexi DUQUE
Norway / Norvège	Øyvind LIER
South Africa / Afrique du Sud	Louis HATTINGH
Spain/ Espagne	Manuel G. DE MEMBRILLERA
USA / États-Unis	Amanda SUTTER

CORRESPONDING MEMBERS

Australia / Australie	Ian LANDON-JONES (until 2016)

INACTIVE MEMBERS / MEMBRES INACTIFS

China / Chine	Sanda YU
Korea / Corée	Ki-Seog KIM

SOMMAIRE	CONTENTS

TABLE DES MATIÈRES

TABLE OF CONTENTS

TABLEAUX & FIGURES

TABLES & FIGURES

TABLES

FIGURES

À LA MÉMOIRE DE DR CHRIS OOSTHUIZEN

Quelques semaines après avoir terminé le premier jet de ce bulletin, notre président et ami Chris Oosthuizen nous quittait le samedi 11 novembre 2017 à l'âge de 70 ans. Chris a été le maître d'œuvre de ce bulletin et, malgré sa lutte de 5 années contre le cancer, il a fait des efforts considérables pour diriger notre Comité et terminer ce bulletin, en particulier durant l'année 2017.

Chris Oosthuizen avait plus de quarante années d'expérience dans l'ingénierie des barrages. Il a mis en place un programme d'évaluation de la sécurité pour plus de 300 barrages publics en Afrique du Sud. Il a développé une méthodologie probabiliste pour l'évaluation de la sécurité des barrages en vue de la priorisation des travaux de réhabilitation à mener. Il a aussi été un expert en sécurité des barrages dans de nombreux pays, tels la Suisse, le Soudan, l'Égypte, l'Éthiopie, la Malaisie, l'Australie, le Mozambique, la Namibie et le Lesotho.

Chris a été un membre très actif du Comité Sud-africain des barrages SANCOLD et de la CIGB. Il a présenté de nombreux articles et animé de nombreuses conférences. Au-delà d'être membre du Comité technique de la surveillance des barrages de la CIGB, lequel a produit les bulletins 138 et 158, il a également contribué à plusieurs autres comités. Depuis 2012, Chris a présidé ce Comité et a été le principal auteur de ce bulletin sur les leçons tirées d'études de cas pour la surveillance des barrages.

Chris était universellement reconnu dans la profession. Il était passionné par la formation professionnelle et le tutorat à tous les niveaux de qualification des personnels techniques. Il a contribué de manière prépondérante à la définition et la mise en œuvre de bonnes pratiques dans la surveillance des barrages et la sécurité en formant des générations de personnels techniques, tant ingénieurs, techniciens, opérateurs et responsables, à quelque niveau que ce soit, de la surveillance des barrages.

Les membres du Comité technique de la surveillance des barrages ont eu la chance de partager et de travailler avec Chris et garderont en mémoire non seulement sa grande connaissance et son immense savoir, mais également sa grande modestie et son caractère chaleureux. Mais le trait de caractère qui les marquera le plus reste son écoute et la capacité à comprendre ses interlocuteurs.

Ce fut l'idée de Chris de dédier ce bulletin à ses modèles Elmo DiBiagio et Ralph Peck. Chris a aussi été un modèle pour tous les membres de ce comité et ce bulletin est également dédié à sa mémoire.

IN MEMORY OF DR. CHRIS OOSTHUIZEN

A few weeks after finishing the final draft of this bulletin, our chairman and friend Chris Oosthuizen passed away on Saturday 11th of November at the age of 70. Chris was the main driver of this bulletin and despite his 5-year battle with cancer he made huge efforts to steer our Committee and to elaborate and conclude the bulletin work, especially during 2017.

Chris Oosthuizen was working in dam engineering for more than four decades. He established a Dam Safety Evaluation Programme for the more than 300 public dams in South Africa and developed a probabilistic methodology for the prioritisation of dam safety rehabilitation works. He also served as Dam Safety Expert in many countries such as Switzerland, Sudan, Egypt, Ethiopia, Malaysia, Australia, Mozambique, Namibia and Lesotho.

Chris was very active in SANCOLD and ICOLD, presenting numerous papers and lectures. Not only was he a member of the ICOLD Technical Committee on Dam Surveillance that produced Bulletins 138 and 158, he also contributed to a number of other technical committees through the years. Since 2012, Chris chaired this ICOLD committee and was the principal author of this latest bulletin of significant surveillance case histories.

Chris was universally respected by all his peers and had an outstanding passion for the professional development and mentoring of all levels of technical staff including dam operators. He contributed in a very important way to good practice in dam surveillance and safety by training generations of technical staff including engineers, technicians, operators as well as those responsible for all aspects of surveillance.

The TCDS Committee members are fortunate to have shared and worked with Chris and will always remember his profound knowledge and professional wisdom together with his great humility and his warm character. The most profound character was however his ability to listen to other people.

It was Chris' idea to dedicate this bulletin to his role models Elmo Dibiagio and Ralph Peck. Chris himself is also a role model and so this bulletin is likewise dedicated to his memory.

PRÉFACE

DÉDICACE

Elmo DiBagio et ipso facto Ralph Peck ont été des modèles pour une grande partie des membres du Comité technique de la surveillance des barrages de la CIGB (TCDS, selon l'acronyme anglais). Lorsque le Comité travaillait sur la structure détaillée du bulletin, Elmo adressait le message suivant aux deux co-présidents :

Je joins à ce message une synthèse d'une page – une des 22 que je viens de finaliser pour une conférence que je donnerai à Chicago l'été prochain en l'honneur de Ralph B. Peck. Les 22 études de cas sont des synthèses sur une seule page d'exemples qui illustrent l'évolution du travail d'instrumentation à NGI au cours des 60 dernières années. J'ignore si vous partagez avec le professeur Peck le concept de "synthèse sur une seule page". Il expliquait à ses étudiants qu'aucun sujet n'est trop compliqué pour ne pas pouvoir être résumé en une seule page ! Il n'aurait accepté selon son enseignement que des contributions tenant sur une seule page.

Le concept de la synthèse sur une seule page a orienté la mise en œuvre de ce bulletin. Pour cette raison, le Comité a considéré approprié de dédier ce bulletin à Elmo DiBiagio et à Ralph Peck, qui est décédé en 2008. Il est par conséquent approprié de reprendre la vision initiale de Ralph Peck sur l'instrumentation.

LA VISION DE RALPH PECK SUR L'INSTRUMENTATION DES BARRAGES

La vision de Ralph Peck sur l'instrumentation des barrages est résumée ci-après. Ce résumé est considéré comme une base essentielle pour ce bulletin (tel que cela apparait dans la préface que Ralph Peck a rédigée pour Dunnicliff & Green en 1988).

L'instrumentation n'est pas une fin en soi

Malgré son rôle essentiel, l'instrumentation ne garantit pas une bonne conception ni une construction sans problèmes ou déficiences. Des instruments inadéquats à de mauvais emplacements ne peuvent qu'au mieux apporter une information qui sèmera le trouble, et au pire détourner l'attention d'autres signaux indicateurs de problèmes. Si l'instrumentation n'est pas une fin en soi, il en va de même pour la sophistication et l'automatisation. Les deux exigences essentielles sont : une sensibilité suffisante pour apporter l'information nécessaire, et la fiabilité pour assurer que les données essentielles soient disponibles durant toute la période utile. Sur ce point, les systèmes sophistiqués et complexes ne présentent pas tous la même fiabilité.

PREFACE

DEDICATION

Elmo Dibiagio and ipso facto Ralph Peck have been role models for several members of the ICOLD Technical Committee on Dam Surveillance (TCDS). When the detail layout of this Bulletin was investigated, Elmo sent the following e-mail to the two co-chairmen:

I have attached a one-page summary one of 22 that I just compiled for a lecture I will give in Chicago next summer at a conference honouring Ralph B. Peck. All 22 case histories are one-page examples that illustrate chronologically the evolution of instrumentation work at NGI during the past 60 years. I don't know if you associate the expression "One-page summary" with Professor Peck. He told his students that no project is so complicated that it cannot be summarized on one folio sheet! He would only accept one-page assignments in the particular course he taught!

The one-page summaries swing the course of action of the bulletin, and it was therefore only considered appropriate to dedicate this bulletin to Elmo Dibiagio and the late Ralph Peck, who died in 2008. It is therefore appropriate to revisit Ralph Peck's views on instrumentation on the onset.

RALPH PECK'S VIEWS ON DAM INSTRUMENTATION

The following is a summary of Ralph Peck's views on instrumentation, which is considered essential as background for this bulletin (as it appeared in a preface, he had written for Dunnicliff & Green in 1988).

Instrumentation is not an end in itself

Notwithstanding its vital role, instrumentation cannot guarantee good design or trouble-free construction. The wrong instruments in wrong places provide information that may at best be confusing, and at worst divert attention from tell-tale signs of trouble. Not only instrumentation is not an end in itself, but neither is sophistication or automation. The two prime requirements are: sufficient sensitivity to provide the necessary information, and reliability to ensure that dependable data can be obtained throughout the needed period. Not all sophisticated systems are equally reliable.

Tous les instruments installés sur un ouvrage doivent être choisis et placés afin de répondre à une question spécifique

Ce principe simple est le garant d'une instrumentation réussie. Il est hélas plus facile d'installer des instruments et de collecter les mesures et seulement ensuite se demander à quelles questions ces mesures pourraient apporter des réponses. L'ingénieur doit non seulement envisager la manière dont l'ouvrage se comportera, pour autant que l'information dont il dispose soit intrinsèquement correcte, mais également les impacts que les lacunes et les insuffisances. Alors seulement il pourra identifier les éléments spécifiques qui permettront soit de confirmer que l'ouvrage se comporte conformément aux hypothèses considérées lors de la conception soit, si ce n'est pas le cas, d'identifier en quoi le comportement diffère substantiellement du comportement prévu. À partir de là, les questions-clés peuvent être formulées et l'ingénieur peut déterminer quels instruments, placés à quels endroits, peuvent apporter les réponses à ces questions. Les progrès de la technologie de surveillance des barrages atteignent une ampleur et une complexité qui vont bien au-delà des exigences et des besoins de l'ingénieur. Cependant, quelles que soient ces applications, chaque instrument devrait être choisi et localisé pour apporter une réponse à une question spécifique. C'est un dogme répandu, par exemple, que chaque barrage en remblai devrait être instrumenté dans l'espoir que les mesures mettront en évidence un défaut insoupçonné et révèleront une éventuelle défaillance ou rupture.

Le système d'auscultation n'a de valeur que s'il est correctement installé

Les instruments d'auscultation sont le plus souvent installés par des techniciens compétents, dans des conditions difficiles. Ils tentent de faire un travail de précision alors qu'ils interviennent sur un chantier en interrompant les travaux ou l'exploitation, et en travaillant parfois la nuit pour essayer de limiter ces interruptions. L'engagement et la persévérance sont le prix du succès, et on ne retrouve généralement pas ces qualités chez les soumissionnaires au prix les plus bas. De plus, l'installateur sera peu motivé par le choix d'instruments de qualité inférieure susceptibles de tomber en panne prématurément ou de fournir des mesures douteuses. Des instruments robustes et fiables ne sont pas nécessairement coûteux, mais un coût bas est rarement un critère de sélection judicieux. Un système d'auscultation ne peut être mis en œuvre avec succès si le critère coût prévaut sur la qualité des instruments et sur l'expérience de l'installateur.

Les instruments en eux-mêmes sont des singularités

Les corps étrangers, tels que les capteurs, introduits dans le système sol/roche/structure, constituent des singularités. Leur présence peut modifier les valeurs mêmes qu'ils sont censés mesurer. L'ingénieur doit en tenir compte dans la conception. Une 'instrumentation trop fournie constitue un gaspillage inutile, alors qu'une instrumentation trop pauvre, par souci de réduction des coûts, peut mener à de fausses économies et peut être préjudiciable. Enfin, le nombre d'instruments doit être suffisant, non seulement pour surmonter les pertes inévitables résultant d'un dysfonctionnement et des dommages causés par les activités de construction, mais également pour donner une représentation significative de la dispersion inhérente des résultats. Le principe de la redondance dans le suivi des grandeurs les plus pertinentes reste fondamental

La valeur des opérateurs d'auscultation et du personnel de maintenance est sous-estimée

L'instrumentation doit être considérée à son juste niveau. Elle constitue un élément de l'activité, plus vaste, de la surveillance. La colonne vertébrale de la surveillance est constituée de personnes correctement formées pour la maintenance, la mesure, le traitement, l'interprétation et l'analyse de l'information. Mais l'observation visuelle, par le meilleur instrument qui soit, un œil averti, apporte le plus souvent toute l'information nécessaire et restera toujours l'élément clé dans la surveillance d'un ouvrage, quel qu'il soit.

Every instrument installed on a project should be selected and placed to assist in answering a specific question

Following this simple rule is the key to successful field instrumentation. Unfortunately, it is easier to install instruments, collect the readings, and then wonder if there are any questions to which the results might provide an answer. The engineer should judge not only the way the design will perform if the information is essentially correct, but how the gaps or shortcomings might influence the performance of the project. Then, and only then, can specific items be identified that will reveal whether the project is performing in accordance with design assumptions, or, if not, in what significant way the performance differs. Thereafter, the critical question can be phrased, and the engineer can determine what instruments, at what locations, would answer those questions. Advancements in surveillance technology are of an extent and complexity far beyond the requirements of the practicing engineer. Yet, in all these applications every instrument should be selected and located to assist in answering a specific question. It is a widely held dogma, for instance, that every earth dam should be instrumented, in the hope that some unsuspected defect will reveal itself in the observations and give warning of an impending failure.

The monitoring system is only as good as the installation

These jobs are usually performed by diligent installers under difficult and unpleasant conditions, trying to do precision work while surrounded by workers whose teamwork or operation of equipment is being interrupted, or working the graveyard shift in an attempt to reduce such interruptions. Dedication of this sort is the price of success, and it is rarely found at the price of the lowest bidder. Moreover, the installer can hardly be motivated to be dedicated to the task of installing instruments of inferior quality that are likely to fail prematurely or to produce questionable data. Rugged reliable instruments are not necessarily expensive, but lowest cost is rarely a valid reason for its choice. No arrangement for a program of instrumentation is a candidate for success if it sets cost above quality of instruments or fee above experience of the installer.

Instruments are discontinuities

Non-representative objects, such as sensors, introduced into the soil/rock/structure systems are discontinuities. Their presence may alter the very quantities they are intended to measure. The designer must take this into account in the design. Too much instrumentation is wasteful and may disillusion those who pay the bills, whilst too little, arising from a desire to save money, can be more than false economy: it can be dangerous. Finally, there must be enough instruments not only to allow for the inevitable losses resulting from malfunction and damage by construction activities, but also to provide a meaningful picture of the inherent scatter in results.

Under-estimated value of instrumentation observers and maintenance staff

Instrumentation needs to be kept in perspective. It is one part of the broader activities of surveillance. Trained people in the maintenance, observation, processing and interpretation as well as analyses remain the backbone of surveillance, Visual observations using the best of all instruments, the human eye, can often provide all the necessary information and are always an essential part of the field observation on any project.

APPROCHE RETENUE POUR CE BULLETIN

Ce bulletin étant amené à constituer un document de référence, les modalités suivantes ont été adoptées :

- La version imprimée du bulletin contient plus de 71 pages d'études de cas provenant de 22 pays. Un lien numérique vers une base de données plus complète sur les cas présentés est fourni, avec des références choisies pour plus d'information. Une mise en page standardisée a été utilisée pour les études de cas (éléments techniques, description de l'incident et enseignements) ;

- Il est prévu de mettre à jour et d'étendre cette base d'information, afin d'intégrer des études de cas d'autres pays, en particulier de Russie, d'Inde, de Chine, d'Australie et de Nouvelle Zélande. De tous ces cas, il sera possible de tirer des enseignements de grande valeur.

REMERCIEMENTS

Le Comité adresse un mot de remerciement à tous ceux qui ont soumis une étude de cas, ainsi qu'à tous ceux qui ont contribué à la mise en forme des études de cas (les éditeurs voudraient en particulier remercier à cet égard Louis Hattingh, Peter Nightingale, María Casasola et Stefan Hoppe).

CHRIS OOSTHUIZEN ET JÜRGEN FLEITZ

**Co-Présidents du Comité Technique de la
Surveillance des Barrages**

Août 2017

APPROACHED USED FOR THIS BULLETIN

With their role models in mind, it is obvious that a minimalistic approach has been taken on this bulletin, making it unique in more ways than one:

- The printed version of the bulletin will contain more than 71 one-page case histories from 22 countries, with a digital link to the slightly more comprehensive versions of the case histories with selected references for more information. A standard format has been used for these case histories (i.e. technical background, description of the incident and lessons learnt).

- Provision has been made for future revisions, in order to include case histories from countries such as Russia, India, China, Australia and New Zealand, all of which can make a major contribution to lessons learnt.

ACKNOWLEDGEMENTS

A word of thanks for everyone who submitted case histories and those assisted with the editing of case histories (the editors would however like to single out Louis Hattingh, Peter Nightingale, María Casasola and Stefan Hoppe for their contribution in this regard).

CHRIS OOSTHUIZEN AND JÜRGEN FLEITZ

**Chairmen of ICOLD Committee on
Dam Surveillance**

August 2017

1. INTRODUCTION

1.1. MANDAT

Les termes de référence (TDR) du Comité technique de la surveillance des barrages (acronyme en anglais TCDS) ont été approuvés à l'occasion de l'Assemblée générale de Kyoto en 2012 pour la période 2013–2016. Ces termes de référence consistent essentiellement à préparer une série de bulletins sur la surveillance des barrages, couvrant :

- Les méthodes d'amélioration de la qualité et de la fiabilité de l'information ;

- Le traitement des données et les techniques de représentation ;

- Le diagnostic efficace pour la détermination des schémas de comportement ;

- Les systèmes de surveillance dédiés pour l'optimisation des coûts de maintenance, de réhabilitation et autres coûts du cycle de vie ;

- Et, l'impact de la surveillance sur la prévention des incidents et des ruptures de barrages, par le biais d'une sélection d'études de cas.

Une réunion informelle des membres du Comité technique TCDS s'est tenue le 4 juin 2012 à Kyoto, au cours de laquelle il fut décidé de commencer par un bulletin sur des études de cas sur la surveillance des barrages. A Stavanger en 2015, les termes de référence ont été prorogés jusqu'en 2018.

1.2. DÉFINITIONS

Afin d'éviter toute confusion de terminologie dans ce document, il a été convenu de retenir les définitions de la surveillance et de l'auscultation utilisées dans les précédents bulletins de la CIGB traitant de la surveillance des barrages, lesquelles sont illustrées à la Figure 1.

Il existe une certaine confusion sur des termes tels qu' « analyse » et « évaluation ». Les définitions suivantes sont utilisées dans ce document. [1]

Analyse : il s'agit du processus de segmentation d'un sujet ou d'un ensemble complexe en composants plus simples dans le but d'en obtenir une meilleure compréhension.

Évaluation : il s'agit du processus de compréhension objective de l'état d'un objet, obtenu par l'observation ou la mesure. En d'autres termes, il s'agit de la description systématique de l'état, de la qualité et de la valeur d'un objet, faisant appel à des critères d'évaluation (déterminés par des normes ou des règles). Le but est d'acquérir une bonne compréhension des actions et décisions antérieures afin de faciliter la réflexion et contribuer à l'identification de futurs changements. On utilise en anglais le terme « assessment ».

Évaluation : il s'agit aussi du processus de revue de l'information, généralement sous forme de données quantifiées, dans le but de juger de sa pertinence et de qualifier sa valeur, soit par comparaison avec des éléments similaires, soit par comparaison avec des normes ou règles. Ce processus est utilisé dans la prise de décision concernant l'adoption, le rejet ou la révision d'un programme ou d'une action. On utilise en anglais le terme « evaluation ».

---

[1] Il faut noter que l'anglais fait la distinction entre les mots « assessment », « evaluation » et « appraisal », qui sont tous traduits en français par le mot « évaluation »

1. INTRODUCTION

1.1. MANDATE

The Terms of Reference (ToR) of the Technical Committee on Dam Surveillance (TCDS) were approved at the Annual Meeting in Kyoto 2012 for the term 2013–2016. The Terms of Reference (ToR) of the present Dam Surveillance Committee are basically to prepare a series of dedicated dam surveillance bulletins covering:

- Methods for the improvement of the quality and reliability of information.

- Data processing and representation techniques.

- Effective Diagnostic analyses to determine behaviour patterns.

- Dedicated surveillance systems for the optimization of maintenance-, rehabilitation- and other life cycle costs; and

- Impact of surveillance (preventing dam incidents and dam failure by means of selected case histories).

An informal meeting of previous TCDS members was held in Kyoto, on 4 June 2012. It was decided to start off with a bulletin on dam surveillance case histories. In Stavanger, the term of the TCDS was extended until 2018.

1.2. DEFINITIONS

In order to avoid any conflict with the various definitions of terminology in this document, it has been decided to follow the definitions of surveillance and monitoring used by previous ICOLD technical committees on dam surveillance, which are graphically represented in Figure 1.

Confusion reigns over some terms such as "analysis", "assessment", "evaluation" and "appraisal". The following relevant definitions are used in this document:

Analysis: is the process of breaking down a complex topic or substance into smaller parts to gain a better understanding of it.

Assessment: is the process of objectively understanding the state or condition of something, by observation and measurement, in other words, a systematic determination of a subject's merit, worth and significance, using evaluation criteria (governed by standards). The purpose is gaining insight into prior or existing initiatives, to enable reflection and assist in the identification of future change.

Evaluation: refers to the process of documenting relevant information, usually in measurable terms and measuring something for the purpose of judging it and of determining its "value", either by comparison to similar things or to a standard. Evaluations therefore often utilise assessment data along with other resources to make decisions about revising, adopting, or rejecting a course of action or program.

Appraisal: is basically an "evaluation" that compares options to deliver a specific objective.

DOI: 10.1201/9781003274841-1

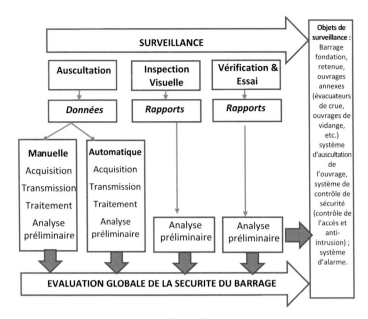

Figure 1
Illustration de la "Surveillance", telle que définie par la CIGB (voir bulletins CIGB 118 et 158)

1.3.   PARAMÈTRES PRINCIPAUX D'AUSCULTATION

Les paramètres principaux d'auscultation/surveillance peuvent être classés dans trois catégories principales :

- Les paramètres associés aux charges (externes et internes) ;

- Les paramètres de réponse ;

- Et les paramètres d'intégrité (parfois aussi appelés paramètres d'état).

Paramètres de charge

Les paramètres de charges peuvent être subdivisés comme suit :

- Charges statiques (gravité et sédimentation) ;

- Charges hydrostatiques (niveaux d'eau), lesquelles sont aussi directement liées aux pressions interstitielles, qui elles sont clairement des paramètres de réponse ;

- Charges cycliques (journalières, saisonnières, annuelles et/ou à long terme) – les effets gravitaires de la lune et du soleil, ainsi que le climat (température, pluie et humidité) sont les plus importants ;

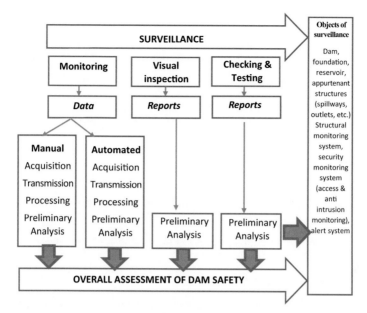

Figure 1.
Graphical representation of "Surveillance" as defined by ICOLD (see ICOLD Bulletins 118 and 158)

1.3.   BASIC PARAMETERS TO BE MONITORED

Monitoring/surveillance parameters can be classified into three main types for discussion purposes, vis-à-vis:

- Load parameters (both external and internal loads);

- Response parameters; and

- Integrity parameters (also referred to as condition parameters).

Load parameters

Load parameters can be subdivided into:

- Static loads (gravity and sedimentation).

- Hydro-static loads (water level) is also directly linked to interstitial water pressure that is strictly speaking a response parameter.

- Cyclic loads (daily, seasonal, annual and/or long-term) - the gravity effect of the moon and sun as well as the climate (temperature, rain, and humidity) are the major contributors.

- Charges dynamiques de faible ou forte intensité. Les charges dynamiques de faible intensité sont par exemple les sollicitations ambiantes naturelles, les sollicitations forcées ou intermittentes, tels que le vent, les véhicules, les piétons, l'action des vagues, les vibrations induites par une centrales hydroélectrique, des organes hydrauliques d'évacuation et des séismes distants. Les charges dynamiques de forte intensité sont principalement causées par des séismes ou des explosions ;

- Charges imprévisibles ou de nature humaine (telles des changements environnementaux majeurs, des actes de guerre, sabotage, vandalisme, minage et excavations en souterrain).

Paramètres de réponse

Les paramètres de réponse aux charges statiques, cycliques et dynamiques sont en particulier :

- Les déplacements (déformations et déplacements mesurés en des points choisis, dans une, deux ou 3 dimensions) ;

- Les rotations (mesurées sur un, deux ou trois axes en des points choisis) ;

- Les infiltrations d'eau, provoquant le développement de pressions interstitielles (aussi appelées sous-pressions lorsqu'elles apparaissent dans la fondation), ainsi que les débits de fuite ou d'infiltration, qui conduisent à des paramètres de réponse secondaires tel le degré de saturation, la densité, la vitesse de propagation d'ondes sonores, la température, la résistivité, etc.

- Les vibrations de faible intensité (amplitude de quelques microns) et de grande intensité (amplitude de quelques mm).

Paramètres d'intégrité

Les paramètres d'intégrité font référence aux paramètres affectant la fonctionnalité du barrage proprement dit (fondation, propriétés du remblai et du béton), les équipements hydromécaniques et électromécaniques, les ouvrages annexes, les ouvrages d'adduction, la retenue et son voisinage. Ces modifications des propriétés matérielles et géophysiques ou leur détérioration sont causées par l'érosion, la cavitation, la fatigue et les réactions chimiques, telles que la réaction alcali-granulat, la carbonatation (sur les surfaces en béton) et la corrosion (des pièces en acier et des armatures des bétons). Une attention particulière sera portée à l'amélioration du suivi de ces paramètres d'intégrité qui sont rarement surveillés dès le début.

1.4. INTERPRÉTATION DES DONNÉES D'AUSCULTATION

Un volume important de mesures est généralement collecté et ces données sont représentées graphiquement en fonction du temps. Toutefois, ces <u>données</u> d'auscultation ne deviennent réellement utiles que lorsqu'elles sont transformées en <u>information</u>. Cela peut se faire, par exemple sous forme de représentations graphiques d'évolution d'un paramètre en fonction du temps ou en fonction d'un autre paramètre. En étudiant ces graphiques, tableaux, etc. d'informations de surveillance, les informations sont transformées en connaissance qui contribue, en considérant l'évolution dans le temps, à la compréhension du comportement du barrage. Le processus est représenté graphiquement à la Figure 2.

- Dynamic loads of low or high intensity. Low intensity dynamic loads are for example natural ambient, forced ambient or intermittent ambient loads, such as wind, vehicles, pedestrians, wave action, induced vibration from the hydropower station, outlet works and distant earthquakes. High intensity dynamic loads are mainly caused by earthquakes and explosives and

- Unforeseen and "man-made" loads (such as drastic environmental changes, acts of war, sabotage, vandalism, blasting and tunneling).

Response parameters

Response parameters from static, cyclic and dynamic load parameters are typically:

- Translations (deformations and displacements measured at selected points in either one, two or three dimensions).

- Rotations (measured around one, two or three axes at selected points).

- Water infiltration causing interstitial pressure that presents itself as pore pressure (and so-called "uplift") as well as seepage and leakage which will induce secondary response parameters such as saturation, density, sound velocity, temperature, resistivity etc. and

- Vibrations of low intensity (amplitudes of several microns) and high intensity (amplitudes of several millimeters).

Integrity parameters

Integrity parameters refer to those parameters affecting the functionality of the dam wall (foundation, soil and concrete properties) as well as mechanical/electrical equipment, appurtenant works, conveyances, the reservoir and its surroundings. These changes in material and geo-physical properties or deterioration are caused by erosion, cavitation and chemical reactions, such as swelling and carbonation (on the surface). Attention will have to be given to improve these measurements as it is seldom monitored from the onset.

1.4. INTERPRETATION OF MONITORING DATA

Massive amounts of data are usually being generated and plotted versus time. However, monitoring data becomes only useful when it is transformed into information. This can, for example, be done by converting 'readings-versus-time' plots to plots of one parameter versus another. By studying these monitoring information graphs, tables, etc., the information is transformed into knowledge that with time evolves into insight in the behaviour of the dam. The process is graphically displayed in Figure 2.

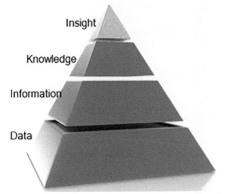

Données – Information – Connaissance - Compréhension

Figure 2
Représentation graphique du processus de conversion depuis la donnée jusqu'à la compréhension du comportement du barrage

La compréhension du comportement peut également être obtenue par la modélisation numérique, en particulier lorsque des modes de défaillance sont identifiés et simulés.

1.5. BULLETINS PRÉCÉDENTS TRAITANT DE LA SURVEILLANCE DES BARRAGES

Les ingénieurs en charge de la sécurité des barrages doivent avoir une connaissance approfondie de l'instrumentation. Les bulletins suivants de la CIGB traitent spécifiquement de la surveillance des barrages :

Bulletin 60 (CIGB : 1988) Auscultation des barrages, considérations générales

Le bulletin 60 est une mise à jour et la fusion des bulletins 21 (1969) et 23 (1972). Ce bulletin est structuré en sept chapitres et une annexe, avec les titres suivants : Introduction, Objectifs de l'auscultation, Conception du dispositif de mesure, Mesures et appareils, Modalités de mise en œuvre en fonction des buts poursuivis et des types de barrages, Fréquence des mesures, Considérations générales sur l'installation des appareils de mesures et Méthodes géodésiques pour la détermination des mouvements des barrages.

Bulletin 68 (CIGB : 1989) Auscultation des barrages et de leurs fondations, la technique actuelle

Le bulletin 68 est constitué de 11 rapports de Comités nationaux, avec une note de synthèse comprenant les titres principaux : Sécurité des barrages, Grandeurs physiques à mesurer et appareils de mesure, Dispositifs de mesure, Auscultation sismique, Fréquence des mesures, Automatisation du dispositif d'auscultation, Analyse des résultats de mesures, Vieux barrages ou barrages sans dispositif adéquat d'auscultation ou avec problèmes particuliers, Nouveaux appareils, Barrages cités dans les rapports nationaux (environ 60 barrages), et Conclusions.

Figure 2.
Graphical representation of the process to convert data into insight of the behaviour of the dam

Further insight can be obtained by numerical modelling, especially if failure modes can be identified and simulated.

1.5. PREVIOUS SURVEILLANCE BULLETINS

Dam surveillance engineers must have more than casual knowledge of instrumentation. The following are the relevant ICOLD bulletins produced on surveillance:

Bulletin 60 (ICOLD: 1988) Dam monitoring, General considerations

Bulletin 60 was an update and merger of Bulletins 21 (1969) and 23 (1972). The bulletin is arranged in seven chapters and an addendum under the following headings: Introduction, Monitoring purposes, Planning instrumentation systems, Measurements and instruments, Installation according to purpose and dam type, Frequency of measurements, General considerations for the installation of instruments and Geodetic methods for the determination of movements of dams.

Bulletin 68 (ICOLD: 1989) Monitoring of dams and their foundations, State of the Art

Bulletin 68 was produced in the format of 11 National Committee Reports with a general overview under the following headings: Dam safety, Physical quantities and related instrumentation, Monitoring systems, Seismic monitoring, Frequency of measurements, Monitoring system automation, Analysis of instrumental data, Older dams or those devoid of an adequate monitoring system or dams with special problems, New instruments, Dams cited in the National Committee Reports (approximately 60 dams) and Conclusions.

Bulletin 87 (CIGB : 1992) Amélioration de l'auscultation des barrages existants

Le bulletin 87 contient des recommandations et des exemples présentés par le Comité ad hoc pour l'auscultation des barrages existants, sous la présidence d'A. Marazio. Le bulletin comporte 12 études de cas provenant de 7 pays.

Bulletin 118 (CIGB : 2000) Systèmes d'auscultation automatique des barrages, recommandations et exemples

Le bulletin 118 est dédié aux solutions d'auscultation automatique des barrages. Il intègre en annexe 12 études de cas provenant de 11 pays.

Bulletin 138 (CIGB : 2009) La surveillance : Élément de base d'un processus "sécurité des barrages"

Le bulletin 138 est une introduction à la surveillance des barrages. Il explique la vision de la CIGB de l'organisation optimale de la surveillance, et décrit tous les éléments nécessaires à la surveillance et à l'auscultation des barrages. Le contenu est de nature <u>conceptuelle</u>, tout en tenant compte de la pratique internationale. Le thème est abordé en termes généraux, faisant référence à des éléments plus détaillés.

Bulletin 158 (CIGB, 2013 pour la version en anglais, 2016 pour la version en français) Guide pour la surveillance des barrages

Le bulletin 158 est un guide pour la surveillance des barrages, d'une approche complémentaire et plus <u>pratique</u> que le bulletin 138. Il fait le point sur les méthodes, les outils et les appareils d'auscultation en indiquant les limitations de leur mise en œuvre et leur complémentarité.

Bulletin 87 (ICOLD: 1992) Improvement of existing dam monitoring

Bulletin 87 contains recommendations and case histories produced by the ad hoc Committee on Monitoring of Existing Dams, with A. Marazio as chairman. The bulletin is in the format of 12 case histories from 7 countries.

Bulletin 118 (ICOLD: 2000) Automated Dam Monitoring Systems, Guidelines and case histories

Bulletin 118 focuses on automated monitoring solutions and includes an annex that covers 12 case histories from 11 countries.

Bulletin 138 (ICOLD: 2009) General Approach to Dam Surveillance

Bulletin 138 provides an optimal organisation of all components required for dam surveillance and monitoring. The contents are more philosophical in nature, to accommodate international state-of-the-art. The subject is generally covered in broad terms with references to detail.

Bulletin 158 (ICOLD: 2013 English and 2016 French) Dam surveillance guide

Bulletin 158 is a guide on dam surveillance, more practical in nature than Bulletin 138, accommodating international state-of-the-art. The subject is generally covered in broad terms with references to detail.

2.   CLASSIFICATION DES ÉTUDES DE CAS ET LEÇONS TIRÉES

Le présent bulletin souhaite montrer la vaste expérience acquise depuis 60 ans par la communauté des ingénieries de barrages dans le domaine de la surveillance des barrages à partir d'études de cas.

Les études de cas documentées (comprenant également les études de cas de référence) visent à couvrir les expériences vécues en relation avec un ou plusieurs aspects des termes de référence du présent bulletin (items a. à e. ci-dessous), ainsi qu'un thème complémentaire qui concerne les exemples de systèmes de gestion de la surveillance des barrages (item f.). Les cas d'études se répartissent ainsi selon les catégories suivantes :

1. Méthodes d'amélioration de la qualité et de la fiabilité de l'information.

2. Traitement des données et techniques de représentation ;

3. Diagnostic efficace pour déterminer les schémas de comportement ;

4. Systèmes de surveillance dédiés pour optimiser les coûts de maintenance, de réhabilitation et les autres coûts du cycle de vie ;

5. Impacts de la surveillance sur la prévention des incidents et ruptures de barrages ;

6. Revue des systèmes de gestion de la surveillance des barrages.

Les études de cas couvrent un large éventail d'aspects techniques et présentent des réussites mais aussi des incidents, dont certains ont eu des conséquences catastrophiques. Le cadre temporel s'étend sur 70 ans, depuis la seconde guerre mondiale jusqu'à aujourd'hui. Le contenu des études de cas relève de la seule responsabilité de leurs auteurs.

Les êtres humains se sont développés parce qu'ils ont appris de leurs erreurs, comme l'a déclaré Rabindranath Tagore "si vous fermez la porte à toutes les erreurs, la vérité sera exclue". Le but est d'apprendre de ces expériences pratiques, et non de critiquer les individus impliqués, qui travaillaient avec les techniques et les pratiques de leur temps.

Ce bulletin souhaite contribuer à poursuivre le partage d'expérience de la communauté des ingénieries de barrages, en particulier dans le domaine de la surveillance des barrages.

Certaines des études de cas sont largement connues et ont été décrites et analysées dans de nombreuses publications. La plupart concernent des ruptures de barrages connues et ont été rassemblées dans ce bulletin en tant qu'études de cas de référence pour mettre l'accent sur les enseignements tirés en matière de surveillance de barrages.

Les neuf études de cas de référence sont les suivantes :

• Barrage de Malpasset (importance de la surveillance des aspects géologiques);

• Barrage du Vajont (importance de l'auscultation des versants de la retenue) ;

• Barrage de Zeuzier (effet extrême de la diminution des pressions interstitielles) ;

• Barrage de Teton (importance de disposer d'observations visuelles appropriées) ;

• Barrages de Dnieprostoi, de Möhne et de l'Eder (bombardements et explosifs pendant la seconde guerre mondiale) ;

2.   CLASSIFICATION OF CASE HISTORIES AND LESSONS LEARNT

This Bulletin attempts to show the vast experience gained over the past 6 decades by the dam engineering community in the field of dam surveillance by means of case histories.

The documented case histories (including the benchmark histories) endeavours to cover the practical experiences related with one or several aspects of the terms of reference (i.e. bullets a. to e. below) as well as an added category to cover examples of dam surveillance management systems (bullet f). The categories used for the case histories are thus as follows:

1.   Methods for the improvement of the quality and reliability of information.

2.   Data processing and representation techniques.

3.   Effective Diagnostic analyses to determine behaviour patterns.

4.   Dedicated surveillance systems for the optimization of maintenance, rehabilitation and other life cycle costs.

5.   Impact of surveillance on preventing dam incidents and dam failure; and

6.   Overview of dam surveillance management systems.

The case histories cover a wide variety of technical aspects and deal with success stories but also incidents, some of them with catastrophic consequences. The time framework spans over 70 years: from the times of the Second World War up to the present. The content of the case histories is solely the responsibility of the individual authors of the case histories.

Human beings have developed because they have learned from their errors, as stated by Rabindranath Tagore: *"If you shut the door to all errors, truth will be shut out."* The purpose is to learn from these practical experiences, not to criticize the involved individuals, who had to work with the techniques and rules of practice available at the time.

This bulletin wants to contribute to keep learning from the experience of the dam engineering community, specifically in the field of dam surveillance.

Some of the case histories are widely known and have been described and analysed in numerous publications. Most of them are related to famous dam failure incidents and were compiled as international "benchmark case histories" for this bulletin in order to focus on the specific lessons learnt related with dam surveillance.

The nine benchmark case histories are the following:

*   Malpasset Dam (importance of monitoring engineering geological aspects);

*   Vajont Dam (importance of monitoring reservoir slopes);

*   Zeuzier Dam (the unbelievable effect of pore pressure relief);

*   Teton Dam (value of diligent visual observations);

*   Dnieprostroi, Möhne and Eder Dams (explosive loads during World War II);

DOI: 10.1201/9781003274841-2

- Barrage de Folsom (rupture de vanne, périodiquement essayée mais seulement partiellement) ;

- Barrage de Cahora Bassa (intérêt d'une installation appropriée pour la durée de vie des appareils d'auscultation) ;

- Barrage de Zoeknog (rupture annoncée par les capteurs de pression interstitielle dont les mesures ont été ignorées) ;

- Barrage de Tous (défaillance des systèmes de secours).

En complément de ces neuf études de cas, les membres et observateurs du comité technique sur la surveillance des barrages ont ajouté 71 autres études, qui représentent un large éventail des aspects pratiques qu'un ingénieur spécialiste de la surveillance est susceptible de rencontrer dans l'exercice de sa profession. Près de 80 ingénieurs exerçant dans le domaine des barrages ont contribué en étant auteurs ou co-auteurs des études de cas de ce bulletin.

Un des principaux objectifs de ce bulletin est de diffuser la large expérience rassemblée dans ces études de cas qui ont été établies par un grand nombre d'ingénieurs depuis plusieurs décennies.

Le présent chapitre cherche à classer les expériences très diverses et hétérogènes des études de cas pour faciliter le choix par le lecteur des exemples les plus pertinents.

Chaque étude de cas fait l'objet d'une fiche-résumé qui figure dans l'addenda de ce bulletin. L'étude de cas complète est présentée dans l'annexe informatique. L'addenda contient également des tableaux avec les informations complémentaires suivantes sur chaque étude de cas :

- objectif principal ;

- intérêt principal ;

- observations ;

- aléas et modes de défaillance ;

- mots-clés ;

- auteurs.

Le tableau 1 résume les 9 études de cas de référence ainsi que les 71 autres études de cas par pays, nom, type et hauteur de barrage, année de mise en service ainsi que les catégories d'études de cas tel que spécifié au chapitre 2.5.

Une première analyse des 80 études de cas (9 études de cas de référence et 71 études de cas complémentaires) montre que :

- 74 études de cas concernent un barrage ou une retenue spécifiques (8 études de cas de référence et 66 études de cas complémentaires) ;

- 2 études de cas concernent un barrage sans que son nom ou son âge soient spécifiés ;

- 2 études de cas concernent deux barrages ou plus (une étude de cas de référence concernant les barrages de Dnieprostri en URSS, de Möhne et d'Eder en Allemagne, et une autre étude de cas complémentaire (barrages de La Minilla et d'El Gergal en Espagne)) ;

- Folsom Dam (gate failure, tested regularly but not all the way through);

- Cahora Bassa Dam (the value of diligent installations on the life of instruments);

- Zoeknog Dam (failure, predicted by pore pressure gauges, but ignored); and

- Tous Dam (backup systems failure).

Further to these 9 benchmark case histories, members and observers of the Technical Committee on Dam Surveillance added a total of 71 additional case histories, representing a wide range of practical aspects a surveillance engineer might encounter in his professional activity. Around 80 professional dam engineers have contributed by means of being authors or co-authors to the case histories of this bulletin.

One of the main purposes of this bulletin is to disseminate the broad experience condensed into the case histories that were made by many dam engineers throughout decades.

The present chapter tries to classify the very diverse and heterogenic experiences of the case histories to facilitate the reader the selection of those examples more directly related to the subject of their interest.

A one-page summary of each case history is also included in the addenda of this bulletin. The full case history is part of the digital appendix. The addenda also contain tables with further information about each case history, such as:

- main objective;

- main benefit;

- observations;

- hazards and failure modes;

- key words; and

- authors.

Table 1 summarise the 9 benchmark histories as well as all 71 case histories by country, name, type and height of dam, the year of commissioning, as well as the case history categories.

A first analysis of the 80 case histories (9 benchmark and 71 other) reveals:

- 74 case histories deal with one specific dam or reservoir (8 benchmark and 66 other);

- 2 case histories deal with a single dam, without specifying the name and age of the dam;

- 2 case history deal with two or more specific dams (1 benchmark (Dnieprostri Dam in the USSR and Möhne and Eder Dams in Germany) and 1 other (La Minilla and El Gergal in Spain));

- 3 barrages font chacun l'objet de deux études de cas (barrage de WAC Bennett au Canada, barrage de Porce II en Colombie et barrage de l'Oued El Makhazine au Maroc parmi les études de cas complémentaires) ;

- 4 études de cas concernent un parc de barrages (exemples en Italie, au Portugal, en Suède et en Espagne parmi les études de cas complémentaires).

Une revue plus approfondie fait le lien entre les études de cas et le sujet principal, quelques études de cas relevant de plusieurs aspects :

- 23 études de cas concernent la fondation (3 études de cas de référence et 20 études de cas complémentaires) ;

- 34 études de cas concernent le corps de barrages en remblai (3 études de cas de référence et 31 études de cas complémentaires) ;

- 17 études de cas concernent le corps de barrages en béton (2 études de cas de référence et 15 études de cas complémentaires) ;

- 7 études de cas concernent la retenue ou des zones à l'aval du barrage (2 études de cas de référence et 5 études de cas complémentaires) ;

- 5 études de cas concernent l'amélioration des dispositifs d'auscultation ou des solutions d'analyse des données efficaces (études de cas complémentaires).

La liste des 76 barrages faisant l'objet des études de cas est présentée dans le tableau 2 suivant (11 barrages font l'objet d'études de cas de référence et 65 barrages d'études de cas complémentaires). Le lecteur notera que les noms de deux barrages au Canada ont été retirés et que les noms de 74 barrages seulement sont connus.

Il est important de rappeler que les 9 études de cas de référence et les 71 études de cas complémentaires ne représentent qu'un faible échantillon si l'on considère le nombre total de barrages existants, qui comprend 58 519 barrages selon le Registre Mondial des Barrages de la CIGB. C'est pourquoi l'« analyse statistique » appliquée aux exemples présentés ne peut pas être considérée comme représentative pour l'ensemble des barrages dans le monde. Le but de ce chapitre est d'extraire des exemples les aspects techniques principaux et les autres caractéristiques pour faciliter l'identification des études de cas qui pourraient être les plus intéressantes pour le lecteur du bulletin.

Plusieurs critères ont été choisis pour classer les 80 études de cas (9 études de cas de référence et 71 études de cas complémentaires) :

- Pays ;

- Type de structure ;

- Age des barrages (depuis la mise en service) ;

- Hauteur des barrages ;

- Catégories en relation avec les termes de référence ;

- Aléas et modes de défaillance.

- 3 dams have each two case histories (WAC Bennett Dam in Canada, Porce II Dam in Colombia and Oued El Makhazine Dam in Morocco) – none of these are benchmark histories; and

- 4 case histories deal with a dam portfolio (examples from Italy, Portugal, Sweden and Spain) – none of these are benchmark histories.

A further review relates the examples with its principal object (some case histories cover more than one aspect):

- 23 case histories refer to foundation aspects (3 benchmark and 20 other);

- 34 case histories refer to embankment dam bodies (3 benchmark and 31 other);

- 17 case histories refer to concrete dam bodies (2 benchmark and 15 other);

- 7 case histories refer to reservoir or downstream river areas (2 benchmark and 5 other); and

- 5 case histories refer to improvement of monitoring systems or to diligent data analysis solutions (no benchmark case histories).

The list of the 76 individual dams treated in the case histories is included in the following Table 2 (11 dams in the benchmark case histories and 65 dams in the other case histories). It is important to note that the names of 2 dams from Canada have been withheld so only the names of 74 dams are known.

It is important to recall the fact that the 9 benchmark case histories and the other 71 case histories are a small sample considering the total population of existing dams, which encompasses 58 519 dams according to the ICOLD World Register of Dams. Therefore the "statistical analysis" applied to the presented examples, cannot be considered representative for all the dams in the world. The purpose of this chapter is to extract main technical aspects and other features from the examples to facilitate the identification of cases that could be of most interest for the reader of the bulletin.

Several criteria have been chosen to classify the 80 case histories (9 benchmark and 71 other):

- Countries

- Type of structures

- Age of dams (since beginning of operation)

- Height of dams

- Categories related with the terms of reference

- Hazards and failure modes

Tableau 1
Liste des études de cas

N°	Pays	Barrage	Type	Description	Mise en service	Hauteur (m)	Catégorie d'étude de cas
Études de cas de référence							
1	France	Malpasset	voûte	double courbure en béton	1954	67	e, b, f
2	Italie	Vajont	voûte	double courbure en béton	1960	262	e, a, b
3	Suisse	Zeuzier	voûte	double courbure en béton	1957	156	e, a, b, c
4	Etats-Unis	Teton	remblai	terre zonée à noyau central en argile	1976	123	e
5	URSS / Allemagne	Dnieprostri / Möhne / Eder	3 poids	béton, maçonnerie, maçonnerie	1932–1913-1914	60, 40, 48	e
6	Etats-Unis	Folsom	poids, remblai	béton, terre zonée à noyau central en limon	1956	104	c, d, e
7	Mozambique	Cahora Bassa	voûte	double courbure en béton	1974	171	a, e
8	Afrique du Sud	Zoeknog	remblai	terre zonée à noyau central en argile	1993	40	e
9	Espagne	Tous	remblai	enrochement et noyau en argile	1978	80	e
Études de cas complémentaires							
1	Argentine	El Chocón	remblai	terre zonée à noyau central en argile	1973	86	c, d, e
2	Autriche	Durlassboden	remblai	terre zonée à noyau central en argile	1968	80	c, d
3	Autriche	Gmuend	poids	béton	1945	37	a
4	Autriche	Zillergründl	voûte	double courbure en béton	1987	186	a, b, c, d
5	Burkina Faso	Comoé	remblai	terre homogène	1991	24	b, c, e
6	Cameroun	Song Loulou	remblai, évacuateur poids	terre, béton	1981	27	c, e
7	Canada	Canal du Kootenay (bief amont)	2 remblais	enrochement à masque amont en béton	1976	27	b, c
8	Canada	WAC Bennett (analyse de tendances)	remblai	terre zonée à noyau central en argile	1968	180	b

(Continued)

Table 1
List of case histories

N°	Country	Dam	Dam Type	Description	Commissioning	Dam height (m)	Case history category
Benchmark histories							
1	France	Malpasset	Arch	Double curvature concrete	1954	67	e, b, f
2	Italy	Vajont	Arch	Double curvature concrete	1960	262	e, a, b
3	Switzerland	Zeuzier	Arch	Double curvature concrete	1957	156	e, a, b, c
4	USA	Teton	Embankment	Zoned earthfill with central clay core	1976	123	e
5	USSR / Germany	Dnieprostri / Möhne / Eder	Gravity / gravity / gravity	Concrete / masonry / masonry	1932-1913-1914	60, 40, 48	e
6	USA	Folsom	Gravity & embankment	Concrete & zoned earthfill with central silty core	1956	104	c, d, e
7	Mozambique	Cahora Bassa	Arch	Double curvature concrete	1974	171	a, e
8	South Africa	Zoeknog	Embankment	Zoned earthfill with central clay core	1993	40	e
9	Spain	Tous	Embankment	Rockfill with clay core	1978	80	e
Case histories							
1	Argentina	El Chocón	Embankment	Zoned earthfill with central clay core	1973	86	c, d, e
2	Austria	Durlassboden	Embankment	Zoned earthfill with central clay core	1968	80	c, d
3	Austria	Gmuend	Gravity	Concrete	1945	37	a
4	Austria	Zillergründl	Arch	Double curvature concrete	1987	186	a, b, c, d
5	Burkina Faso	Comoé	Embankment	Homogeneous earthfill	1991	24	b, c, e
6	Cameroun	Song Loulou	Embankment with gravity spillway	Earthfill & concrete	1981	27	c, e
7	Canada	Kootenay Canal Forebay	2 embankments	Both concrete faced rockfill	1976	27	b, c
8	Canada	WAC Bennett (trends analysis)	Embankment	Zoned earthfill with central clay core	1968	180	b

(Continued)

N°	Pays	Barrage	Type	Description	Mise en service	Hauteur (m)	Catégorie d'étude de cas
9	Canada	WAC Bennett (robot sous-marin)	remblai	terre zonée à noyau central en argile	1968	180	c, d
10	Canada	non communiqué	remblai	sable et graviers avec noyau central en till	non mentionné	94,5	a,b
11	Canada	non communiqué	remblai	sable et graviers avec noyau central en till	non mentionné	94,5	c,e
12	Colombie	Porce II (études d'évaluation)	poids, remblai	BCR, terre homogène	2001	118	c, d, e
13	Colombie	Porce II (seuils)	poids, remblai	BCR, terre homogène	2001	118	c,e
14	Colombie	Santa Rita	3 remblais	terre à noyau en limon (barrage principal)	1976	51,5, 32, 47	e, c, d
15	Colombie	Tona	remblai	enrochement à masque amont en béton	2015	103	a, c
16	Colombie	Urrá I	remblai	terre à noyau central en argile	2000	73	e
17	République Tchèque	Mšeno	poids	maçonnerie	1908	20	a, d, e
18	Egypte	El-Karm	poids	béton	1998	25	e
19	France	Etang	remblai	terre et enrochement avec membrane PVC amont	1980	33,5	d, e
20	France	Grand'Maison	remblai	enrochement à noyau central en argile	1985	160	a, d, e
21	France	La Palière	remblai	terre homogène	1982	6,5	c, d
22	France	Mirgenbach	remblai	terre homogène	1983	22	a, b, c, e
23	Allemagne	Sylvenstein	remblai	terre zonée à noyau central en argile	1959	48	d, e
24	Iran	Alborz	remblai	enrochement à noyau central en argile	2009	78	a, d
25	Iran	Gotvand	remblai	enrochement zoné à noyau central en argile	2011	182	a, b, f
26	Iran	Karun IV	voûte	double courbure en béton	2010	230	b, c
27	Iran	Masjed-e-Soleiman	remblai	enrochement à noyau central en argile	2002	177	a, b

(Continued)

N°	Country	Dam	Dam Type	Description	Commissioning	Dam height (m)	Case history category
9	Canada	WAC Bennett (ROV)	Embankment	Zoned earthfill with central clay core	1968	180	c, d
10	Canada	Not reported	Embankment	Sand/gravel fill with central till core	Not reported	94,5	a,b
11	Canada	Not reported	Embankment	Sand/gravel fill with central till core	Not reported	94,5	c,e
12	Colombia	Porce II (assessment study)	Gravity & embankment	RCC & homogeneous earthfill	2001	118	c, d, e
13	Colombia	Porce II (thresholds)	Gravity & embankment	RCC & homogeneous earthfill	2001	118	c,e
14	Colombia	Santa Rita	3 embankments	Earthfill – silty core at main dam	1976	51,5, 32, 47	e, c, d
15	Colombia	Tona	Embankment	Concrete faced rockfill	2015	103	a, c
16	Colombia	Urrá I	Embankment	Earthfill with clay core	2000	73	e
17	Czech Republic	Mšeno	Gravity	Masonry	1908	20	a, d, e
18	Egypt	El-Karm	Gravity	Concrete	1998	25	e
19	France	Etang	Embankment	Zoned earthfill and rockfill with upstream PVC lining	1980	33,5	d, e
20	France	Grand'Maison	Embankment	Rockfill with central clay core	1985	160	a, d, e
21	France	La Palière	Embankment	Homogeneous earthfill	1982	6,5	c, d
22	France	Mirgenbach	Embankment	Homogeneous earthfill	1983	22	a, b, c, e
23	Germany	Sylvenstein	Embankment	Zoned earthfill with central clay core	1959	48	d, e
24	Iran	Alborz	Embankment	Rockfill with central clay core	2009	78	a, d
25	Iran	Gotvand	Embankment	Zoned rockfill with central clay core	2011	182	a, b, f
26	Iran	Karun IV	Arch	Double curvature concrete	2010	230	b, c
27	Iran	Masjed-e-Soleiman	Embankment	Rockfill with central clay core	2002	177	a, b

(Continued)

N°	Pays	Barrage	Type	Description	Mise en service	Hauteur (m)	Catégorie d'étude de cas
28	Iran	Seymareh	voûte	double courbure en béton	2011	180	a, b
29	Italie	Ambiesta	voûte	double courbure en béton	1960	58,6	e, c, d
30	Italie	Isola Serafini	contreforts, évacuateur poids	béton	1962	32,5	e, c, d
31	Italie	San Giacomo	contreforts	béton	1950	97,5	e, c, d
32	Italie	MISTRAL	-	-	-	-	f
33	Japon	Retenue supérieure de Kyogoku	remblai	enrochement à masque amont en béton bitumineux	2013	59,9	a, d, e
34	Japon	Okuniikappu	voûte	double courbure en béton	1963	61,2	a
35	Japon	Yashio	remblai	enrochement à masque amont en béton bitumineux	1992	90,5	c, e
36	Maroc	Oued El Makhazine (parafouille)	remblai	enrochement à noyau central en argile	1978	66,5	b, c, d
37	Maroc	Oued El Makhazine (joint en galerie)	remblai	enrochement à noyau central en argile	1978	66,5	b, c, d
38	Maroc	Tuizgui Ramz	poids	maçonnerie	2007	24,5	c
39	Norvège	Muravatn	remblai	enrochement à noyau central en till	1968	77	d, e
40	Norvège	Storvatn	remblai	enrochement à noyau central incliné en béton bitumineux	1987	90	c, d
41	Norvège	Svartevann	remblai	enrochement à noyau central en till	1976	129	c, d, e
42	Norvège	barrage test	remblai	enrochement à étanchéité centrale par membrane bitumineuse	1969	12	c, d
43	Norvège	Viddalsvatn	remblai	enrochement à noyau central en moraine glaciaire	1972	96	c, d, e

(Continued)

N°	Country	Dam	Dam Type	Description	Commissioning	Dam height (m)	Case history category
28	Iran	Seymareh	Arch	Double curvature concrete	2011	180	a, b
29	Italia	Ambiesta	Arch	Double curvature concrete	1960	58,6	e, c, d
30	Italia	Isola Serafini	Buttress/gravity weir	Concrete	1962	32,5	e, c, d
31	Italia	San Giacomo	Buttress	Concrete	1950	97,5	e, c, d
32	Italia	MISTRAL	-	-	-	-	f
33	Japan	Retenue supérieure de Kyogoku	Embankment	Asphalt faced rockfill	2013	59,9	a, d, e
34	Japan	Okuniikappu	Arch	Double curvature concrete	1963	61,2	a
35	Japan	Yashio	Embankment	Asphalt faced rockfill	1992	90,5	c, e
36	Morocco	Oued El Makhazine (cutoff wall)	Embankment	Rockfill with central clay core	1978	66,5	b, c, d
37	Maroc	Oued El Makhazine (culvert joint)	Embankment	Rockfill with central clay core	1978	66,5	b, c, d
38	Maroc	Tuizgui Ramz	Gravity	Masonry	2007	24,5	c
39	Norway	Muravatn	Embankment	Rockfill with central moraine till core	1968	77	d, e
40	Norway	Storvatn	Embankment	Rockfill with central inclined asphaltic concrete core	1987	90	c, d
41	Norway	Svartevann	Embankment	Rockfill with central moraine till core	1976	129	c, d, e
42	Norway	Trial dam	Embankment	Rockfill with central bitumen membrane	1969	12	c, d
43	Norway	Viddalsvatn	Embankment	Rockfill with central glacial moraine core	1972	96	c, d, e

(*Continued*)

N°	Pays	Barrage	Type	Description	Mise en service	Hauteur (m)	Catégorie d'étude de cas
44	Norvège	Zelazny Most (Pologne) gestBarragens	stériles	forme circulaire	1975	22–60	e
45	Portugal		-	-	-	-	a, b, c, f
46	Roumanie	Gura Râului	contreforts	béton	1979	74	c, e
47	Roumanie	Paltinu	voûte	double courbure en béton	1971	108	e
48	Roumanie	Pecineagu	remblai	enrochement zoné à masque amont en béton	1985	105	b, f
49	Roumanie	Poiana Uzului	contreforts	béton	1970	80	c, e
50	Afrique du Sud	Belfort	voûtes multiples, remblai	béton, terre homogène	1976	17	e
51	Afrique du Sud	Driekoppies	remblai, évacuateur poids	terre zonée avec noyau central en argile, béton	1998	50	a, e
52	Afrique du Sud	Inyaka	remblai, évacuateur à coursier	terre zonée avec noyau central en argile, béton	2001	56	a, b, c, e, f
53	Lesotho	Katse	voûte	double courbure en béton	1996	185	a, b, c, e
54	Afrique du Sud	Kouga	voûte	double courbure en béton	1969	72	a, c, e
55	Afrique du Sud	Ohrigstad	remblai	enrochement à masque amont en béton	1955	52	c
56	Espagne	La Aceña	poids arqué	béton	1991	65	a, c, d, e
57	Espagne	Caspe II	remblai	terre zonée avec noyau central en argile	1987	56	e, c
58	Espagne	Cortes	poids arqué	béton	1988	116	e
59	Espagne	La Minilla, El Gergal	poids voûte	béton	1974	63	d
60	Espagne	La Loteta	remblai	terre zonée avec noyau central et tapis amont en argile	2008	34	e, c

(Continued)

N°	Country	Dam	Dam Type	Description	Commissioning	Dam height (m)	Case history category
44	Norway	Zelazny Most (Poland)	stériles	Ring shaped	1975	22–60	e
45	Portugal	gestBarragens	-	-	-	-	a, b, c, f
46	Romania	Gura Râului	Buttress	Concrete	1979	74	c, e
47	Romania	Paltinu	Arch	Double curvature concrete	1971	108	e
48	Romania	Pecineagu	Embankment	Concrete faced zoned rockfill	1985	105	b, f
49	Romania	Poiana Uzului	Buttress	Round headed concrete	1970	80	c, e
50	South Africa	Belfort	Multiple buttressed arch & embankment	Concrete & homogeneous earthfill	1976	17	e
51	South Africa	Driekoppies	Embankment & gravity spillway	Zoned earthfill with central clay core & concrete	1998	50	a, e
52	South Africa	Inyaka	Embankment & trough spillway	Zoned earthfill with central clay core & concrete	2001	56	a, b, c, e, f
53	Lesotho	Katse	Arch	Double curvature concrete	1996	185	a, b, c, e
54	South Africa	Kouga	Arch	Double curvature concrete	1969	72	a, c, e
55	South Africa	Ohrigstad	Embankment	Concrete faced rockfill	1955	52	c
56	Spain	La Aceña	Arch gravity	Concrete	1991	65	a, c, d, e
57	Spain	Caspe II	Embankment	Zoned earthfill with central clay core	1987	56	e, c
58	Spain	Cortes	Arch gravity	Concrete	1988	116	e
59	Spain	La Minilla, El Gergal	Arch gravity	Concrete	1974	63	d
60	Spain	La Loteta	Embankment	Zoned earthfill with central clay core & upstream clay blanker	2008	34	e, c

(Continued)

65

N°	Pays	Barrage	Type	Description	Mise en service	Hauteur (m)	Catégorie d'étude de cas
61	Espagne	Siurana	poids	béton	1972	63	c, d, e
62	Espagne	Val	poids	BCR	1997	90	b, c, f
63	Espagne	55 barrages du bassin de l'Ebre (Damdata)	-	-	-	-	a, b, f
64	Suède	Storfinnforsen	Remblai	terre zonée avec noyau central en till et palplanches bois	1954	23	b, c, d, e
65	Suède	20 barrages à forts enjeux	-	-	-	-	d
66	Tunisie	Ziatine	Remblai	terre zonée avec noyau central en argile	2012	32,5	e, c, d
67	Etats-Unis	Dorris	Remblai	terre homogène	~1928	8	a, e
68	Etats-Unis	Ochoco	Remblai	terre homogène	1919	46	e
69	Etats-Unis	Steinaker	Remblai	terre zonée avec noyau central en argile	1961	50	e
70	Etats-Unis	Wanapum	Poids et remblai	béton, terre zonée	1963	56	e
71	Etats-Unis	Wolf Creek	Poids et remblai	béton, terre	1951	79	e

N°	Country	Dam	Dam Type	Description	Commissioning	Dam height (m)	Case history category
61	Spain	Siurana	Gravity	Concrete	1972	63	c, d, e
62	Spain	Val	Gravity	RCC	1997	90	b, c, f
63	Spain	55 dams in Ebro basin (DAMDATA)	-	-	-	-	a, b, f
64	Sweden	Storfinnforsen	Embankment	Zoned fill with central glacial till core and timber sheet pile	1954	23	b, c, d, e
65	Sweden	20 hight risk dam	Embankment	-	-	-	d
66	Tunisia	Ziatine	Embankment	Zoned earthfill with central clay core	2012	32,5	e, c, d
67	USA	Dorris	Embankment	Homogeneous earthfill	~1928	8	a, e
68	USA	Ochoco	Embankment	Homogeneous earthfill	1919	46	e
69	USA	Steinaker	Embankment	Zoned earthfill with central clay core	1961	50	e
70	USA	Wanapum	Gravity & embankment	Concrete & zoned earthfill	1963	56	e
71	USA	Wolf Creek	Gravity & embankment	Concrete & earthfill	1951	79	e

Tableau 2
Liste des barrages utilisés dans les études de cas

N°	Pays	Barrage	Type	Description	Mise en service	Hauteur (m)
Études de cas de référence						
1	France	Malpasset	voûte	double courbure en béton	1954	67
2	Italie	Vajont	voûte	double courbure en béton	1960	262
3	Suisse	Zeuzier	voûte	double courbure en béton	1957	156
4	Etats-Unis	Teton	remblai	terre zonée à noyau central en argile	1976	123
5	URSS	Dnieprostri	poids	béton	1932	60
6	Allemagne	Möhne	poids	maçonnerie	1913	40
7	Allemagne	Eder	poids	maçonnerie	1914	48
8	Etats-Unis	Folsom	poids, remblai	béton, terre zonée à noyau central en limon	1956	104
9	Mozambique	Cahora Bassa	voûte	double courbure en béton	1974	171
10	Afrique du Sud	Zoeknog	remblai	terre zonée à noyau central en argile	1993	40
11	Espagne	Tous	remblai	enrochement et noyau en argile	1978	80
Études de cas complémentaires						
1	Argentine	El Chocón	remblai	terre zonée à noyau central en argile	1973	86
2	Autriche	Durlassboden	remblai	terre zonée à noyau central en argile	1968	80
3	Autriche	Gmuend	poids	béton	1945	37
4	Autriche	Zillergründl	voûte	double courbure en béton	1987	186
5	Burkina Faso	Comoé	remblai	terre homogène	1991	24
6	Cameroun	Song Loulou	remblai, évacuateur poids	terre, béton	1981	27
7	Canada	Canal du Kootenay (bief amont)	2 remblais	enrochement à masque amont en béton	1976	27
8	Canada	WAC Bennett	remblai	terre zonée à noyau central en argile	1968	180
9	Canada	*non communiqué*	remblai	sable et graviers avec noyau central en till	non mentionné	94,5

(Continued)

Table 2
List of case histories

N°	Country	Dam	Dam Type	Description	Commissioning	Dam height (m)	Case history category
Benchmark histories							
1	France	Malpasset	Arch	Double curvature concrete	1954	67	e, b, f
2	Italy	Vajont	Arch	Double curvature concrete	1960	262	e, a, b
3	Switzerland	Zeuzier	Arch	Double curvature concrete	1957	156	e, a, b, c
4	USA	Teton	Embankment	Zoned earthfill with central clay core	1976	123	e
5	USSR / Germany	Dnieprostri / Möhne / Eder	Gravity / gravity / gravity	Concrete / masonry / masonry	1932-1913-1914	60, 40, 48	e
6	USA	Folsom	Gravity & embankment	Concrete & zoned earthfill with central silty core	1956	104	c, d, e
7	Mozambique	Cahora Bassa	Arch	Double curvature concrete	1974	171	a, e
8	South Africa	Zoeknog	Embankment	Zoned earthfill with central clay core	1993	40	e
9	Spain	Tous	Embankment	Rockfill with clay core	1978	80	e
Case histories							
1	Argentina	El Chocón	Embankment	Zoned earthfill with central clay core	1973	86	c, d, e
2	Austria	Durlassboden	Embankment	Zoned earthfill with central clay core	1968	80	c, d
3	Austria	Gmuend	Gravity	Concrete	1945	37	a
4	Austria	Zillergründl	Arch	Double curvature concrete	1987	186	a, b, c, d
5	Burkina Faso	Comoé	Embankment	Homogeneous earthfill	1991	24	b, c, e
6	Cameroun	Song Loulou	Embankment with gravity spillway	Earthfill & concrete	1981	27	c, e
7	Canada	Kootenay Canal Forebay	2 embankments	Both concrete faced rockfill	1976	27	b, c
8	Canada	WAC Bennett (trends analysis)	Embankment	Zoned earthfill with central clay core	1968	180	b
9	Canada	WAC Bennett (ROV)	Embankment	Zoned earthfill with central clay core	1968	180	c, d

(Continued)

N°	Pays	Barrage	Type	Description	Mise en service	Hauteur (m)
10	Canada	*non communiqué*	remblai	sable et graviers avec noyau central en till	non mentionné	94,5
11	Colombie	Porce II	poids, remblai	BCR, terre homogène	2001	118
12	Colombie	Santa Rita	3 remblais	terre à noyau en limon (barrage principal)	1976	51,5, 32, 47
13	Colombie	Tona	remblai	enrochement à masque amont en béton	2015	103
14	Colombie	Urrá I	remblai	terre à noyau central en argile	2000	73
15	République Tchèque	Mšeno	poids	maçonnerie	1908	20
16	Egypte	El-Karm	poids	béton	1998	25
17	France	Etang	remblai	terre et enrochement avec membrane PVC amont	1980	33,5
18	France	Grand'Maison	remblai	enrochement à noyau central en argile	1985	160
19	France	La Palière	remblai	terre homogène	1982	6,5
20	France	Mirgenbach	remblai	terre homogène	1983	22
21	Allemagne	Sylvenstein	remblai	terre zonée à noyau central en argile	1959	48
22	Iran	Alborz	remblai	enrochement à noyau central en argile	2009	78
23	Iran	Gotvand	remblai	enrochement (zoné) à noyau central en argile	2011	182
24	Iran	Karun IV	voûte	double courbure en béton	2010	230
25	Iran	Masjed-e-Soleiman	remblai	enrochement à noyau central en argile	2002	177
26	Iran	Seymareh	voûte	double courbure en béton	2011	180
27	Italie	Ambiesta	voûte	double courbure en béton	1960	58,6
28	Italie	Isola Serafini	contreforts, évacuateur poids	béton	1962	32,5
29	Italie	San Giacomo	contreforts	béton	1950	97,5
30	Japon	Retenue supérieure de Kyogoku	remblai	enrochement à masque amont en béton bitumineux	2013	59,9
31	Japon	Okuniikappu	voûte	double courbure en béton	1963	61,2

(Continued)

N°	Country	Dam	Dam Type	Description	Commissioning	Dam height (m)	Case history category
10	Canada	Not reported	Embankment	Sand/gravel fill with central till core	Not reported	94,5	a,b
11	Canada	Not reported	Embankment	Sand/gravel fill with central till core	Not reported	94,5	c,e
12	Colombia	Porce II (assessment study)	Gravity & embankment	RCC & homogeneous earthfill	2001	118	c, d, e
13	Colombia	Porce II (thresholds)	Gravity & embankment	RCC & homogeneous earthfill	2001	118	c,e
14	Colombia	Santa Rita	3 embankments	Earthfill – silty core at main dam	1976	51,5, 32, 47	e, c, d
15	Colombia	Tona	Embankment	Concrete faced rockfill	2015	103	a, c
16	Colombia	Urrá I	Embankment	Earthfill with clay core	2000	73	e
17	Czech Republic	Mšeno	Gravity	Masonry	1908	20	a, d, e
18	Egypt	El-Karm	Gravity	Concrete	1998	25	e
19	France	Etang	Embankment	Zoned earthfill and rockfill with upstream PVC lining	1980	33,5	d, e
20	France	Grand'Maison	Embankment	Rockfill with central clay core	1985	160	a, d, e
21	France	La Palière	Embankment	Homogeneous earthfill	1982	6,5	c, d
22	France	Mirgenbach	Embankment	Homogeneous earthfill	1983	22	a, b, c, e
23	Germany	Sylvenstein	Embankment	Zoned earthfill with central clay core	1959	48	d, e
24	Iran	Alborz	Embankment	Rockfill with central clay core	2009	78	a, d
25	Iran	Gotvand	Embankment	Zoned rockfill with central clay core	2011	182	a, b, f
26	Iran	Karun IV	Arch	Double curvature concrete	2010	230	b, c
27	Iran	Masjed-e-Soleiman	Embankment	Rockfill with central clay core	2002	177	a, b
28	Iran	Seymareh	Arch	Double curvature concrete	2011	180	a, b
29	Italia	Ambiesta	Arch	Double curvature concrete	1960	58,6	e, c, d
30	Italia	Isola Serafini	Buttress/gravity weir	Concrete	1962	32,5	e, c, d
31	Italia	San Giacomo	Buttress	Concrete	1950	97,5	e, c, d

(Continued)

N°	Pays	Barrage	Type	Description	Mise en service	Hauteur (m)
32	Japon	Yashio	remblai	enrochement à masque amont en béton bitumineux	1992	90,5
33	Maroc	Oued El Makhazine	remblai	enrochement à noyau central en argile	1978	66,5
34	Maroc	Tuizgui Ramz	poids	maçonnerie	2007	24,5
35	Norvège	Muravatn	remblai	enrochement à noyau central en till	1968	77
36	Norvège	Storvatn	remblai	enrochement à noyau central incliné en béton bitumineux	1987	90
37	Norvège	Svartevann	remblai	enrochement à noyau central en till	1976	129
38	Norvège	barrage test	remblai	enrochement à étanchéité centrale par membrane bitumineuse	1969	12
39	Norvège	Viddalsvatn	remblai	enrochement à noyau central en moraine glaciaire	1972	96
40	Norvège	Zelazny Most (Pologne)	stériles	forme circulaire	1975	22–60
41	Roumanie	Gura Râului	contreforts	béton	1979	74
42	Roumanie	Paltinu	voûte	double courbure en béton	1971	108
43	Roumanie	Pecineagu	remblai	enrochement zoné à masque amont en béton	1985	105
44	Roumanie	Poiana Uzului	contreforts	béton	1970	80
45	Afrique du Sud	Belfort	voûtes multiples, remblai	béton, terre homogène	1976	17
46	Afrique du Sud	Driekoppies	remblai, évacuateur poids	terre zonée avec noyau central en argile, béton	1998	50
47	Afrique du Sud	Inyaka	remblai, évacuateur à coursier	terre zonée avec noyau central en argile, béton	2001	56
48	Lesotho	Katse	voûte	double courbure en béton	1996	185
49	Afrique du Sud	Kouga	voûte	double courbure en béton	1969	72
50	Afrique du Sud	Ohrigstad	remblai	enrochement à masque amont en béton	1955	52

(Continued)

N°	Country	Dam	Dam Type	Description	Commissioning	Dam height (m)	Case history category
32	Italia	MISTRAL	-	-	-	-	f
33	Japan	Retenue supérieure de Kyogoku	Embankment	Asphalt faced rockfill	2013	59,9	a, d, e
34	Japan	Okuniikappu	Arch	Double curvature concrete	1963	61,2	a
35	Japan	Yashio	Embankment	Asphalt faced rockfill	1992	90,5	c, e
36	Morocco	Oued El Makhazine (cutoff wall)	Embankment	Rockfill with central clay core	1978	66,5	b, c, d
37	Maroc	Oued El Makhazine (culvert joint)	Embankment	Rockfill with central clay core	1978	66,5	b, c, d
38	Maroc	Tuizgui Ramz	Gravity	Masonry	2007	24,5	c
39	Norway	Muravatn	Embankment	Rockfill with central moraine till core	1968	77	d, e
40	Norway	Storvatn	Embankment	Rockfill with central inclined asphaltic concrete core	1987	90	c, d
41	Norway	Svartevann	Embankment	Rockfill with central moraine till core	1976	129	c, d, e
42	Norway	Trial dam	Embankment	Rockfill with central bitumen membrane	1969	12	c, d
43	Norway	Viddalsvatn	Embankment	Rockfill with central glacial moraine core	1972	96	c, d, e
44	Norway	Zelazny Most (Poland)	stériles	Ring shaped	1975	22–60	e
45	Portugal	gestBarragens	-	-	-	-	a, b, c, f
46	Romania	Gura Râului	Buttress	Concrete	1979	74	c, e
47	Romania	Paltinu	Arch	Double curvature concrete	1971	108	e
48	Romania	Pecineagu	Embankment	Concrete faced zoned rockfill	1985	105	b, f
49	Romania	Poiana Uzului	Buttress	Round headed concrete	1970	80	c, e
50	South Africa	Belfort	Multiple buttressed arch & embankment	Concrete & homogeneous earthfill	1976	17	e

(Continued)

N°	Pays	Barrage	Type	Description	Mise en service	Hauteur (m)
51	Espagne	La Aceña	poids arqué	béton	1991	65
52	Espagne	Caspe II	remblai	terre zonée avec noyau central en argile	1987	56
53	Espagne	Cortes	poids arqué	béton	1988	116
54	Espagne	El Gergal	poids voûte	béton	1974	63
55	Espagne	La Minilla	poids	béton	1956	70
56	Espagne	La Loteta	remblai	terre zonée avec noyau central et tapis amont en argile	2008	34
57	Espagne	Siurana	poids	béton	1972	63
58	Espagne	Val	poids	BCR	1997	90
59	Suède	Storfinnforsen	Remblai	terre zonée avec noyau central en till et palplanches bois	1954	23
60	Tunisie	Ziatine	Remblai	terre zonée avec noyau central en argile	2012	32,5
61	Etats-Unis	Dorris	Remblai	terre homogène	~1928	8
62	Etats-Unis	Ochoco	Remblai	terre homogène	1919	46
63	Etats-Unis	Steinaker	Remblai	terre zonée avec noyau central en argile	1961	50
64	Etats-Unis	Wanapum	Poids et remblai	béton, terre zonée	1963	56
65	Etats-Unis	Wolf Creek	Poids et remblai	béton, terre	1951	79

N°	Country	Dam	Dam Type	Description	Commissioning	Dam height (m)	Case history category
51	South Africa	Driekoppies	Embankment & gravity spillway	Zoned earthfill with central clay core & concrete	1998	50	a, e
52	South Africa	Inyaka	Embankment & trough spillway	Zoned earthfill with central clay core & concrete	2001	56	a, b, c, e, f
53	Lesotho	Katse	Arch	Double curvature concrete	1996	185	a, b, c, e
54	South Africa	Kouga	Arch	Double curvature concrete	1969	72	a, c, e
55	South Africa	Ohrigstad	Embankment	Concrete faced rockfill	1955	52	c
56	Spain	La Aceña	Arch gravity	Concrete	1991	65	a, c, d, e
57	Spain	Caspe II	Embankment	Zoned earthfill with central clay core	1987	56	e, c
58	Spain	Cortes	Arch gravity	Concrete	1988	116	e
59	Spain	La Minilla, El Gergal	Arch gravity	Concrete	1974	63	d
60	Spain	La Loteta	Embankment	Zoned earthfill with central clay core & upstream clay blanker	2008	34	e, c
61	Spain	Siurana	Gravity	Concrete	1972	63	c, d, e
62	Spain	Val	Gravity	RCC	1997	90	b, c, f
63	Spain	55 dams in Ebro basin (DAMDATA)	-	-	-	-	a, b, f
64	Sweden	Storfinnforsen	Embankment	Zoned fill with central glacial till core and timber sheet pile	1954	23	b, c, d, e
65	Sweden	20 hight risk dam	Embankment	-	-	-	d
66	Tunisia	Ziatine	Embankment	Zoned earthfill with central clay core	2012	32.5	e, c, d
67	USA	Dorris	Embankment	Homogeneous earthfill	~1928	8	a, e
68	USA	Ochoco	Embankment	Homogeneous earthfill	1919	46	e
69	USA	Steinaker	Embankment	Zoned earthfill with central clay core	1961	50	e
70	USA	Wanapum	Gravity & embankment	Concrete & zoned earthfill	1963	56	e
71	USA	Wolf Creek	Gravity & embankment	Concrete & earthfill	1951	79	e

2.1.   ÉTUDES DE CAS PAR PAYS

Comme mentionné ci-avant, le présent bulletin rassemble 9 études de cas de référence et 71 études de cas complémentaires. Les 9 études de cas de référence proviennent de 9 pays (1 étude de cas concerne des barrages de 2 pays, 2 études de cas proviennent des États-Unis). Les 71 études de cas complémentaires proviennent de 22 pays, comme illustré par la carte de la Figure 3 qui montre les pays ayant soumis des études de cas. De même, le nombre d'études de cas par pays est montré sur la Figure 4.

Figure 3
Pays ayant soumis des études de cas

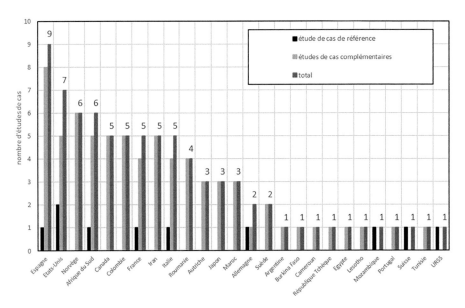

Figure 4
Number of case histories per country

2.1. CASE HISTORIES BY COUNTRIES

As already mentioned, this bulletin contains 9 benchmark case histories, in addition to 71 case history contributions. The 9 benchmark case histories are from 9 different countries (1 case history have dams from 2 different countries while 2 case histories are from the USA). The 71 other case histories are from 22 different countries, shown in Figure 3. Likewise, the number of case histories per country is portrayed in Figure 4. Number of case histories per country.

Figure 3
Map showing the countries that submitted case histories

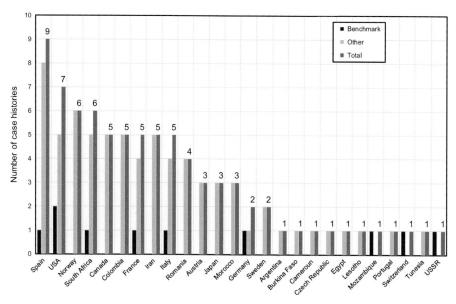

Figure 4
Number of case histories per country

2.2. TYPE DE STRUCTURES

Les types de structures de chaque étude de cas ont été analysés. Il est important de noter que dans quelques cas les barrages relèvent de plusieurs types de structures. Seuls les types de structures abordés dans les études de cas sont fournis. La répartition par type de structure est présentée en Figure5. Il en ressort que :

- La majorité des structures concernées par les études de cas sont des barrages en remblai (41 structures ou 54%) suivis par les barrages poids (12 structures ou 16%) et les voûtes (11 structures ou 14%) ;

- Les barrages en remblai :

  - Sur les 24 barrages en remblai, 16 sont des barrages zonés et 5 sont des barrages homogènes. Les détails ne sont pas connus pour les autres barrages en remblai ;

  - Sur les 16 barrages en enrochements, 4 sont des barrages à masque amont en béton, 2 sont des barrages à masque amont en béton bitumineux et les autres sont munis d'un noyau ;

  - Un barrage en remblai mixte terre/enrochement est muni d'une étanchéité amont par membrane PVC.

- Barrages poids :

  - Sur les 12 barrages poids, 7 sont en béton, 4 en maçonnerie et 1 en béton compacté au rouleau (BCR) ;

- Barrages voûtes :

  - Les 11 barrages voûtes sont des structures à double courbure en béton ;

- Barrages à contreforts :

  - Les 3 barrages de ce type sont tous en béton ;

- Versants de la retenue :

  - 3 études de cas concernent les versants de la retenue.

- Barrages poids voûtes et barrages à voûtes multiples :

  - Un barrage de chaque type, en béton ;

- Autres structures :

  - On compte également un évacuateur de crue, un barrage de stériles et l'association d'un barrage poids en BCR et d'un barrage en remblai homogène.

2.2.   TYPE OF STRUCTURES

For this the type of structure for each of the case histories were analysed. It is important to note that in some cases the dams are a combination of a number of structures. Only those relevant to the case histories are provided. A graphical presentation of the structure type distribution is provided in 5. The following is evident from the analyses:

- The majority of structures discussed in the case histories are embankment dams (41 structures or 54%) followed by gravity dams (12 structures or 16%) and arch dams (11 structures or 14%);

- Embankment dams:

  - Of the 24 earthfill embankments, 16 are zoned and 5 are homogeneous. No further detail is available for the rest.

  - Of the 16 rockfill embankments, 4 are concrete faced, 2 are asphalt faced and the rest all have some or another core.

  - There is also 1 mixed fill embankment with an upstream PVC liner.

- Gravity dams:

  - The 12 gravity dams consist of 7 concrete, 4 masonry and 1 roller compacted concrete (RCC) structures.

- Arch dams:

  - All 11 arch dams are double curvature concrete structures.

- Buttress dams:

  - There is 3 of these types of structures and they are all concrete.

- Reservoir slopes:

  - Three of the case histories deals with reservoir slopes.

- Arch gravity, gravity arch and multiple arch dams:

  - There is 1 each of these types of structures and both are concrete.

- Other:

  - Finally, there is also a weir structure, a tailings dam and a combination RCC gravity and homogeneous earthfill dam.

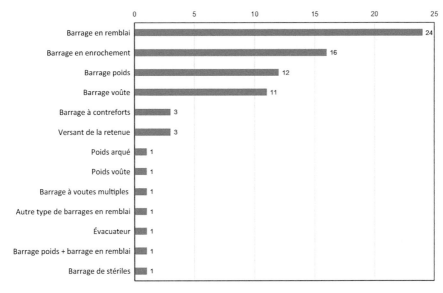

Figure 5
Répartition par type de structure

2.3. AGE DES BARRAGES

L'année 2018 a été retenue comme année de référence pour déterminer l'âge des barrages. En complément, il est important de noter qu'un certain nombre d'études de cas de référence (3 au total, barrage de Malpasset, barrage de Teton et barrage de Zoeknog) concernent une rupture qui n'a pas été suivie par une reconstruction. Pour ces barrages, l'âge a été déterminé comme le nombre d'années entre la mise en service et la rupture. Deux études de cas provenant du Canada concernent un barrage dont le nom et l'âge n'ont pas été communiqués. Ces deux barrages n'ont pas été pris en compte dans l'analyse compte tenu de ce manque d'information.

En référence à l'année de mise en service et au début d'exploitation pour les 74 barrages pour lesquels l'information était disponible, les barrages sont âgés de 110 ans (barrage de Mšeno) à 3 ans seulement (barrage de Tona). Les âges moyen et médian sont respectivement de 41 et 42 ans.

Les âges des barrages sont présentés sur la Figure 7 (sur laquelle les barrages des études de cas de référence figurent en rouge) et la répartition des barrages suivant leur âge (par intervalles de 10 ans) est donnée par la Figure 6. Celle-ci met en évidence les points suivants :

- Compte tenu de l'âge médian de 42 ans, il n'est pas surprenant que la majorité des barrages (16) ait entre 40 et 50 ans ;

- Un nombre considérable de barrages (10) a entre 0 et 10 ans (dont les trois études de cas avec rupture citées ci-avant).

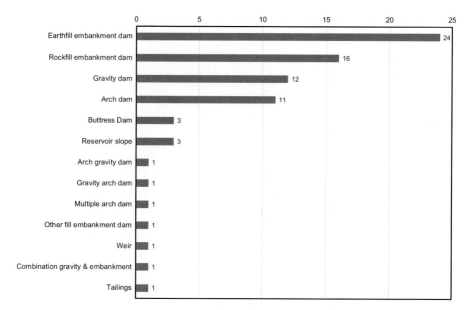

Figure 5
Structure type distribution

2.3. AGE OF DAMS

For the purpose of this exercise the base year to determine the age of each of the dams were taken as 2018. It is also important to note that a number of the benchmark case histories (3 in total – Malpasset Dam, Teton Dam and Zoeknog Dam) deals with the failure of dams that were not subsequent rebuilt. For these dams their age was determined to span from the year of commissioning until the failure. Two case histories from Canada deal with a single without specifying the name and age of the dam. These two dams were not included in this analysis due to a lack of information.

With reference to the year of commissioning and beginning of operation for all 74 dams for which information are available, the ages range from 110 years (Mšeno dam) to only 3 years (Tona dam). The average age is 41 years while the median age is similar at 42 years.

A graphical presentation of each dam's age is given in Figure 7 (with the benchmark histories shown in red) while the distribution of dams according to age (using 10-year intervals) is given in Figure 6. From Figure 6 the following is evident:

- Taking into account the median age of 42 years it is no surprise that the largest number of dams (16 in total) are between 40 and 50 years old; and

- A significant number of dams (10 in total) are between 0 and 10 years old (including the 3 benchmark failures described in detail above).

2.4. HAUTEUR DES BARRAGES

La hauteur des 76 barrages qui font l'objet des études de cas (11 barrages dans les études de cas de référence et 65 barrages dans les études de cas complémentaires) s'étend de 6,5 m (barrage de la Palière) à 262 (barrage du Vajont). Les hauteurs moyenne et médiane sont respectivement de 79 m et 66 m.

Les hauteurs des barrages sont fournies par la Figure 8 (sur laquelle les barrages des études de cas de référence figurent en rouge) et la répartition des barrages suivant leur hauteur (par intervalles de 10 m) est donnée par la Figure 9. Celle-ci met en évidence les points suivants :

- Compte tenu de la hauteur médiane de 66 m, il n'est pas surprenant que la majorité des barrages (9 au total) ait une hauteur entre 50 et 60 m ;

- Un nombre considérable de barrages (5 au total) ont une hauteur entre 180 et 190 m ;

- Trois barrages ne remplissent pas les critères de la CIGB définissant les grands barrages – ces études de cas sont toutefois intéressantes pour le bulletin.

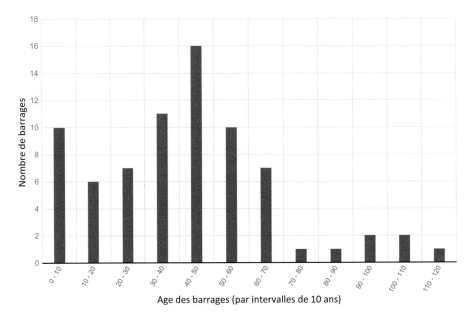

Figure 6
Répartition des barrages suivant leur âge (en années par rapport à 2018)

2.4.   DAM HEIGHT

With regard to the height of the 76 individual dams treated in the case histories (11 dams in the benchmark case histories and 65 dams in the other case histories), it ranges from 6.5 m (La Paliere Dam) to 262 m (Vajont Dam). The average height is 79 m while the median height is only 66 m.

A graphical presentation of the dam heights is given in Figure 8 (with the benchmark histories shown in red) while the distribution of dams according to height (using 10 m intervals) is given in Figure 9. From Figure 9 the following is evident:

- Taking into account the median height of 66 m it is no surprise that the largest number of dams (9 in total) are between 50 and 60 m tall.

- A significant number of dams (5 in total) are between 180 and 190 m tall.

- Three dams actually do not fulfil the ICOLD definition of a large dam – these case histories however are of significant value for the bulletin.

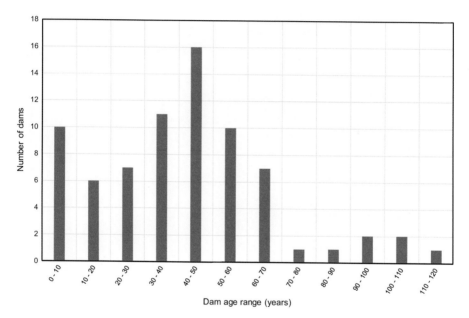

Figure 6
Distribution of dam age (years with 2018 as base year)

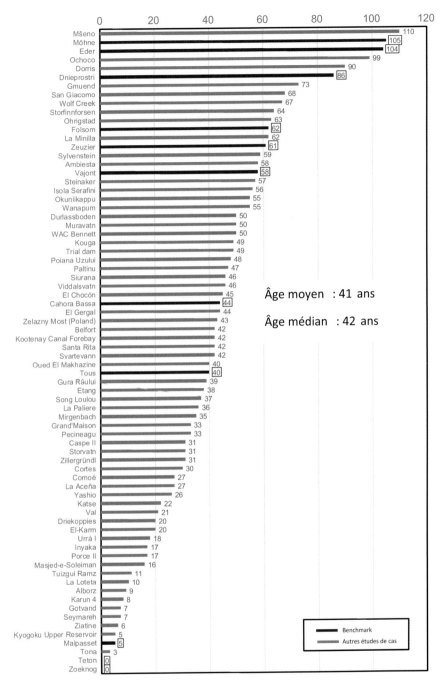

Figure 7
Âge des barrages (en années par rapport à 2018)

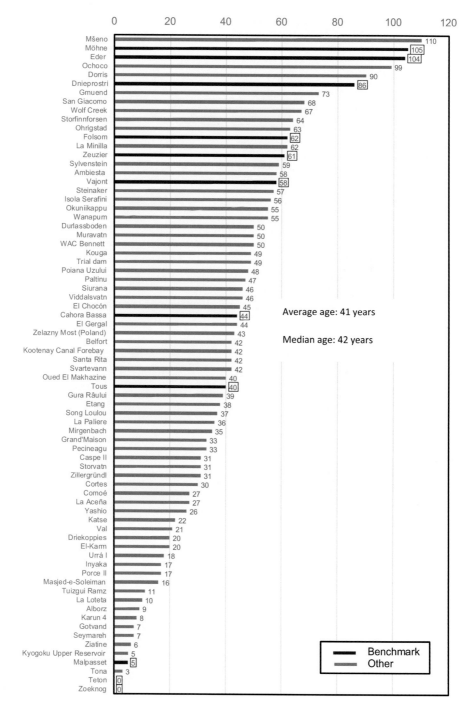

Figure 7
Dam age (years with 2018 as base year)

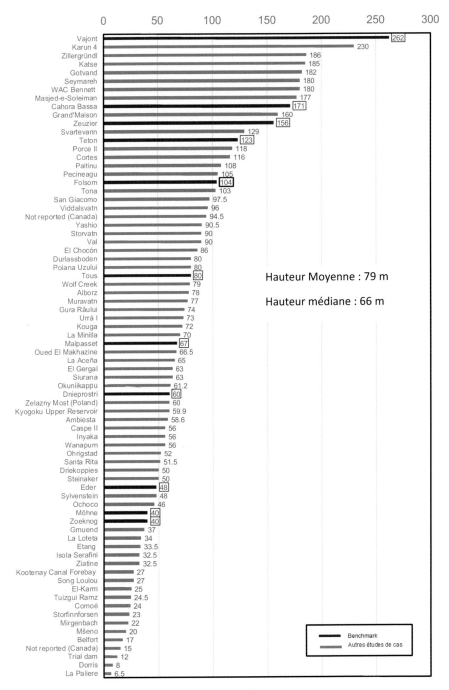

Figure 8
Hauteur des barrages

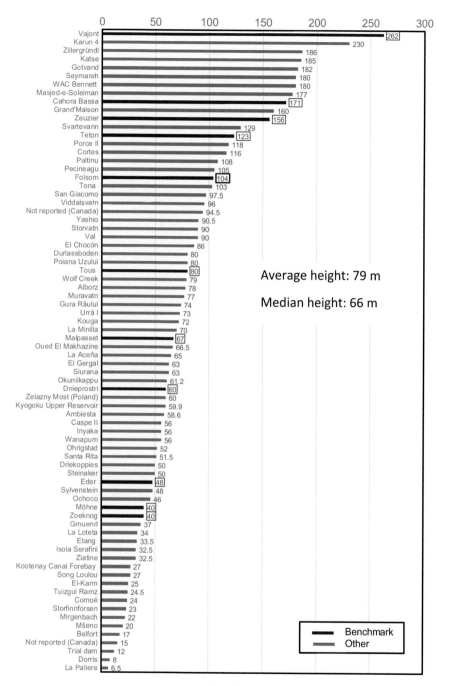

Figure 8
Dam height

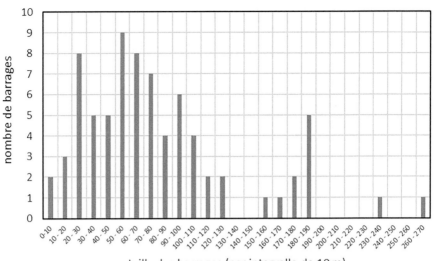

Figure 9
Répartition des barrages suivant leur hauteur

2.5. ÉTUDES DE CAS PAR CATÉGORIES

Chaque étude de cas a été affectée à une ou plusieurs catégories. Une synthèse de la répartition des études de cas par catégorie est fournie dans le Tableau 3, qui met en évidence les points suivants :

- Études de cas de référence (9 études de cas) :

    – La totalité des études de cas de référence relève de la catégorie « (e) impact de la surveillance sur la prévention des incidents et rupture de barrages » alors que seulement 5 relèvent d'au moins une autre catégorie.

- Études de cas complémentaires :

    – La majorité des 71 études de cas complémentaires relève à la fois de la catégorie « (e) impact de la surveillance sur la prévention des incidents et rupture de barrages » (43 études de cas, 60,6%) ainsi que de la catégorie « (c) diagnostic efficace pour déterminer les schémas de comportement » (40 études de cas, 56,3%) ;

    – Une part significative relève de la catégorie « (d) systèmes de surveillance dédiés à l'optimisation des coûts de maintenance, réhabilitation et autres coûts du cycle de vie » (29 études de cas, 40,8%) ;

    – Seules 7 études de cas (9,9%) relèvent de la catégorie « (f) revue des systèmes de gestion de la surveillance des barrages ».

- Total :

    – La majorité des 80 études de cas (51 études de cas, 63,8%) relève de la catégorie « (e) impact de la surveillance sur la prévention des incidents et rupture de barrages » alors que 8 études de cas seulement (10%) concernent la catégorie « (f) revue des systèmes de gestion de la surveillance des barrages ».

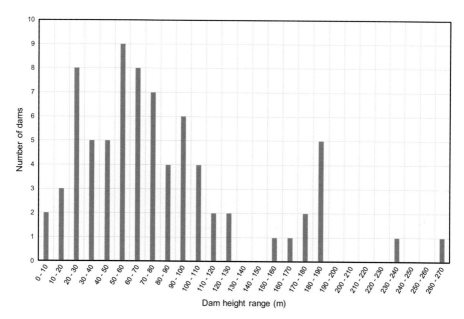

Figure 9
Dam height distribution

2.5. CASE HISTORIES PER CATEGORIES

A case history category or categories have been assigned to each of the case histories. A summary of the distribution of the case history categories is provided in Table 3. The following is evident from this summary:

- Benchmark (9 case histories):

    - All of the 9 benchmark case histories deal with the category "(e) Impact of surveillance on preventing dam incidents and dam failure", whereas only 1 to 3 deal with any of the other categories.

- Other (71 case histories):

    - The majority of the 71 benchmark case histories deal both with the category "(e) Impact of surveillance on preventing dam incidents and dam failure" (43 cases, 60.6%) as well as the category "(c) Effective diagnostic analyses to determine behaviour patterns" (40 cases, 56.3%).

    - A significant portion deals with the category "(d) Dedicated surveillance systems for the optimization of maintenance-, rehabilitation- and other life cycle costs" (29 cases, 40.8%); and

    - Only 7 cases (9.9%) deal with the category "(f) Overview of dam surveillance management systems".

- Total:

    - The majority of the 80 case histories (51 cases, 63.8%), deal with the category "(e) Impact of surveillance on preventing dam incidents and dam failure", whereas only 8 cases (10%) deal with the category "(f) Overview of dam surveillance management systems".

Tableau 3
Répartition des études de cas par catégorie (nombre et pourcentage)

	(a) Méthodes d'amélioration de la qualité et de la fiabilité de l'information	(b) Traitement des données et techniques de représentation	(c) Diagnostic efficace pour déterminer les schémas de comportement	(d) Systèmes de surveillance pour optimiser les coûts de maintenances, réhabilitation et autres coûts du cycle de vie	(e) Impacts de la surveillance sur la prévention des incidents et ruptures de barrages	(f) Revues des systèmes de gestion de la surveillance des barrage
Benchmark	3	3	2	1	9	1
	33.3%	33.3%	22.2%	11.1%	100.0%	11.1%
Études de cas complémentaires	21	19	40	29	43	7
	29.6%	26.8%	56.3%	40.8%	60.6%	9.9%
Total	23	21	41	30	51	8
	28.8%	26.3%	51.3%	37.5%	63.8%	10.0%

Table 3
Case history category distribution (number and %)

	(a) Methods for the improvement of the quality and reliability of information	(b) Data processing and representation techniques	(c) Effective diagnostic analyses to determine behaviour patterns	(d) Dedicated surveillance systems for the optimization of maintenance-, rehabilitation- and other life cycle costs	(e) Impact of surveillance on preventing dam incidents and dam failure	(f) Overview of dam surveillance management systems
Benchmark	3	3	2	1	9	1
	33.3%	33.3%	22.2%	11.1%	100.0%	11.1%
Other	21	19	40	29	43	7
	29.6%	26.8%	56.3%	40.8%	60.6%	9.9%
Total	23	21	41	30	51	8
	28.8%	26.3%	51.3%	37.5%	63.8%	10.0%

2.6. RÉPARTITION DES ÉTUDES DE CAS PAR INITIATEURS DE DÉFAILLANCE OU MODES POTENTIELS DE RUPTURE

Dans son chapitre introductif, le Bulletin 138 de la CIGB indique :

« La surveillance des barrages vise à gérer [le] risque et réduire au mieux sa probabilité d'occurrence, en mobilisant les moyens nécessaires à l'identification précoce d'évènements indésirables susceptibles d'engendrer une éventuelle défaillance ou rupture. Toute organisation d'un processus de surveillance devrait donc viser à faire en sorte que l'on réduise au maximum les probabilités de défaillance par :

- L'identification des modes de rupture et leur prise en compte dans un programme de surveillance ;

- La détection précoce de phénomènes avant-coureurs et évolutifs qui pourraient mener à ces mécanismes de rupture ;

- La connaissance, via des paramètres physiques, du comportement du barrage et de ses composantes ».

Dans ce but, 80 études de cas ont été analysées pour déterminer un ou plusieurs initiateurs de modes de défaillance ou rupture qui ont été perçus par des activités de surveillance. Une liste avec cette information est présentée dans l'addenda avec les fiches-résumés.

De plus, le Tableau 4 rassemble les initiateurs et modes de défaillance ou de rupture qui ont été identifiés. Peu d'études de cas traitent d'amélioration de dispositifs d'auscultation, d'outils efficaces pour l'analyse des mesures d'auscultation, de sabotage ou d'exploitation et maintenance efficaces. Bien que ces aspects ne soient ni des initiateurs ni des modes de défaillance, ils ont été ajoutés à la liste pour compléter la classification des exemples.

Chaque étude de cas peut être rattachée à une ou plusieurs des catégories du Tableau 4. Ce tableau met en évidence les points suivants :

- Étant donné que la majorité des études de cas (55%) concernent des barrages en remblai, il n'est pas surprenant que les problèmes de percolation et d'érosion soient parmi les risques/mode de défaillance ou rupture les plus fréquents ;

- Les pressions interstitielles, les tassements, les déformations et déplacements du corps du barrage, les sous-pressions et le vieillissement du béton et la fissuration comptent également parmi les autres initiateurs/modes de défaillance ou rupture prépondérants.

Une matrice reliant l'initiateur ou le mode de défaillance ou de rupture à chaque étude de cas, ainsi qu'un index des mots-clés, figure dans l'addenda.

2.6. CASE HISTORIES PER HAZARDS OR FAILURE MODES

In its introductory chapter, ICOLD Bulletin 138 defines:

"Dam surveillance aims at managing (…) risk and reducing the probability of occurrence by providing a means of early identification of undesirable events that can possibly cause failure. The organization of any surveillance process should thus aim to reduce the probability of failure as much as possible by:

- *Identification of potential failure modes and providing a surveillance program to cover these,*

- *Early detection of initial stages of evolving phenomena that can lead to failure mechanisms,*

- *Understanding the behaviour of the dam and its components using physical parameters."*

Therefore, all 80 case histories have been analyzed to determine one or several hazards or failure modes that were addressed by the surveillance activities. In this regard, the addenda with the one-page summaries contains a list with this information.

Furthermore, Table 4 contains the hazards and failure modes that were identified. Few cases histories deal with the improvement of monitoring systems or with diligent monitoring data analysis tools or sabotage or proper operation and maintenance. Despite the fact that they are neither hazards nor failure modes, these aspects were added to the list in order to complete the classification of the examples.

Each case history may be related with one or several of the categories in Table 4. From the table the following are evident:

- Given the fact that the majority of the case histories (55%) are related to embankment dams, it is not surprising that seepage and erosion problems are among the most frequent hazards/failure modes; and

- Other prevalent hazards/failure modes also include pore water pressures, settlements, deformations and movements of dam body, uplift and ageing of concrete & cracking.

A matrix which relates the hazard or failure mode with each case history, as well as a key word index, is included in the addenda.

Tableau 4
Répartition des études de cas en fonction des initiateurs et des modes potentiels de
rupture ou de défaillance

Initiateur ou mode potentiel de rupture/défaillance	Études de cas					
	Benchmark		Complémentaires		Total	
Infiltration	1	11.1%	25	35.2%	26	32.5%
Érosion (fondation et corps du barrages)	1	11.1%	15	21.1%	16	20.0%
Pressions interstitielles	2	22.2%	8	11.3%	10	12.5%
Fracturation hydraulique	1	11.1%	4	5.6%	5	6.3%
Déformation de la fondation rocheuse	1	11.1%	5	7.0%	6	7.5%
Tassements, déformations et déplacement (corps du barrage)	1	11.1%	9	12.7%	10	12.5%
Sous-pressions	1	11.1%	7	9.9%	8	10.0%
Glissement (corps du barrage)	-	-	1	1.4%	1	1.3%
Chargement thermique	-	-	1	1.4%	1	1.3%
Vieillissement du béton et fissuration	1	11.1%	6	8.5%	7	8.8%
Membranes d'étanchéité (fissures et comportement)	-	-	4	5.6%	4	5.0%
Stabilité des pentes du barrage	-	-	4	5.6%	4	5.0%
Érosion aval de la rivière	-	-	1	1.4%	1	1.3%
Glissement des versants de la retenue	1	11.1%	2	2.8%	3	3.8%
Séisme	-	-	3	4.2%	3	3.8%
Sédimentation	-	-	1	1.4%	1	1.3%
Amélioration des dispositifs d'auscultation	2	22.2%	3	4.2%	5	6.3%
Analyse efficace des mesures d'auscultation	1	11.1%	2	2.8%	3	3.8%
Sabotage	1	11.1%	-	-	1	1.3%
Exploitation et maintenance efficace	2	22.2%	-	-	2	2.5%

2.7. LEÇONS TIRÉES

Des leçons importantes peuvent être tirées de ces études de cas. Elles ont été classées dans ce bulletin en lien avec les termes de référence du présent comité technique pour la surveillance des barrages. Dans le chapitre précédent, elles ont également été classées en termes d'initiateurs et de mode de défaillance.

Quoi qu'il en soit, ces leçons peuvent aussi être classées selon plusieurs autres catégories telles que :

Pour la fondation :

- Géologie de la fondation ;

- Investigations géotechniques (propriétés des matériaux) ;

Table 4
Case histories per hazards and failure modes

Initiator or potential mode of failure / failure	Case histories					
	Benchmark		Other		Total	
Seepage	1	11.1%	25	35.2%	26	32.5%
Erosion (foundation and dam body)	1	11.1%	15	21.1%	16	20.0%
Pore water pressures	2	22.2%	8	11.3%	10	12.5%
Hydraulic fracture	1	11.1%	4	5.6%	5	6.3%
Foundation rock deformation	1	11.1%	5	7.0%	6	7.5%
Settlements, deformations and movements of dam body	1	11.1%	9	12.7%	10	12.5%
Uplift	1	11.1%	7	9.9%	8	10.0%
Sliding of dam body	-	-	1	1.4%	1	1.3%
Temperature load	-	-	1	1.4%	1	1.3%
Ageing of concrete & cracking	1	11.1%	6	8.5%	7	8.8%
Sealing membranes (cracks & behaviour)	-	-	4	5.6%	4	5.0%
Dam slope stability	-	-	4	5.6%	4	5.0%
Downstream river erosion	-	-	1	1.4%	1	1.3%
Reservoir slope sliding	1	11.1%	2	2.8%	3	3.8%
Earthquake	-	-	3	4.2%	3	3.8%
Sedimentation	-	-	1	1.4%	1	1.3%
Improvement of monitoring systems	2	22.2%	3	4.2%	5	6.3%
Diligent monitoring data analysis	1	11.1%	2	2.8%	3	3.8%
Sabotage	1	11.1%	-	-	1	1.3%
Proper operation and maintenance	2	22.2%	-	-	2	2.5%

2.7.   LESSONS LEARNT

Important lessons can be learnt from these case histories. The lessons learnt in this bulletin have been classified in relation to the terms of reference (ToR) of the present dam surveillance technical committee. In the previous chapter the lessons have also been classified in terms of hazard and modes of failure.

However, these lessons can also be classified in several other categories depending on the specific purpose, for example:

For foundations:

- Foundation geology

- Geotechnical investigations (material properties)

- Versants de la retenue (rupture potentielle de pentes) ;

- Variations des pressions interstitielles (dans la fondation, superficielle ou profonde) ;

- Infiltration et érosion.

Pour le barrage :

- Vérification du chargement, de la réponse au chargement et des paramètres d'intégrité structurelle ;

- Infiltration et érosion ;

- Sollicitations dynamiques et réponse aux sollicitations ;

- Intégrité des équipements hydromécaniques et électriques (importance de disposer de programmes d'exploitation et de maintenance réalisables et adaptés).

Des études de cas peuvent être utilisées dans le cadre de formations à la surveillance des barrages et dans un but d'information pour montrer :

- Les conséquences d'une rupture de barrage ;

- Les conséquences si les signaux d'alarme sont négligés ou ignorés ;

- Comment développer une attitude d'apprentissage à partir des faits de manière objective et professionnelle ;

- L'importance d'une équipe de surveillance dotée de ressources et de moyens suffisants

- La chaîne des activités de surveillance ;

- L'importance d'un système de surveillance efficace, de la conception, de la fourniture, du programme de construction, de l'installation, des observations visuelles, de la gestion des données, du traitement des données, des inspections visuelles, de l'évaluation des résultats ainsi que la maintenance et la mise à niveau périodiques du système ;

- L'importance de la préservation de la mémoire institutionnelle.

Les leçons sur ces aspects sont tirées des études de cas de référence et des études de cas complémentaires documentées dans le présent bulletin.

- Reservoir slopes (potential failure slopes)

- Pore pressure variations (in the foundations, shallow as well as deep ones)

- Seepage and erosion

For dam walls:

- Verification of load, response, and structural integrity parameters

- Seepage and erosion

- Dynamic loads and response

- Hydro Mechanical and Electrical equipment integrity (stressing the importance of practical and proper O&M programmes)

For dam surveillance training and information purposes case histories may be selected to demonstrate:

- Consequences of a dam failure

- Consequences if neglected or warnings ignored

- Development of an attitude to learn from the facts in an unbiased and professional manner

- Importance of a well-resourced and supported dam surveillance team

- Chain of activities of dam surveillance

- Value of diligent surveillance system, design, procurement, construction programming, installation, observations, data management, data processing, visual inspections, evaluation of the results as well as regular maintenance and upgrading of the system

- Importance of the retention of institutional memory

Lessons on these aspects are covered in both the benchmark case-histories as well as the documented case histories in the Bulletin.

3. BOITE A OUTILS POUR LE BENCHMARK DES ETUDES DE CAS

3.1. OBJET DU BENCHMARK DES ÉTUDES DE CAS

Les études de cas constituent un élément didactique important pour l'animation de présentations ou de stages de formation. Les leçons tirées des études de cas servent notamment à confirmer un point existant ou mettre en évidence un point nouveau. Les leçons tirées d'une étude de cas particulière peuvent être légèrement adaptées pour les mettre à la portée du public visé :

- Management ;

- Collègues ;

- Mentors ;

- Étudiants ;

- Concepteurs de barrages ;

- Concepteurs de systèmes d'auscultation ;

- Responsables de l'analyse du comportement ;

- Exploitants de barrages ;

- Ouvriers, techniciens …

- Membres d'équipes en charge d'analyse des modes potentiels de rupture, d'études de dangers, … etc.

Les présentations et la sélection des études de cas utilisées par les conférenciers peuvent varier selon le contexte. Les études de cas peuvent être considérées comme une boîte à outils utilisable par toute personne souhaitant communiquer sur le sujet. Ces boîtes à outils sont spécifiques à plus d'un titre :

- Chaque présentateur (formateur …) a des préférences en matière d'études de cas ;

- Le recours à des études de cas particulières tirées de la boîte à outils dépend du public ;

- Les études de cas concernant une région, un pays ou un type de barrages peuvent avoir davantage d'impact quand le public y perçoit des similitudes avec les problèmes qui les concernent.

3.   TOOLBOX OF BENCHMARK CASE HISTORIES

3.1.   PURPOSE OF BENCHMARK CASE HISTORIES

Case histories are an essential element for any successful motivational presentation and for training purposes. The lessons of the case histories are used to reinforce a point or to make a point. The lessons of a particular case history could be slightly adapted to suit the particular audience. The audiences may be quite diverse. Audiences depend on the purpose of the presentation and can typically be:

- Management.

- Colleagues.

- Mentees (or mentors).

- Students.

- Dam designers.

- Monitoring system designers.

- Behaviour analysts.

- Dam operators.

- Labourers; and/or

- Members of a Potential Failure Mode Analyses team (PFMA) etc.

Presentations as well as the set of case histories used by presenters are thus not universal. The so-called "benchmark" case histories can be viewed as individual presenters' toolbox of case histories. These toolboxes are unique in more ways than one:

- Every presenter (trainer, etc.) have their own preferences for case histories.

- The use of particular examples (case histories) in the toolbox depends on the audience; and

- Regional or country related case histories may have much more impact than irrelevant case histories (e.g. case histories for audience of operators without similar dams under their jurisdiction).

DOI: 10.1201/9781003274841-3

3.2. BUTS DES ÉTUDES DE CAS

Les plus importants sont :

- En premier lieu, d'apprendre des faits historiques (par exemple, la conception du barrage, le dispositif d'auscultation ou les décisions prises par le personnel en charge de l'exploitation … etc.) plutôt que de les critiquer ;

- En second lieu, d'insister sur les faits de manière objective et professionnelle sans incriminer quiconque.

De cette manière, des leçons significatives peuvent être tirés de ces études de cas, notamment en ce qui concerne les aspects liés à la conception, à la construction et aux procédures d'exploitation.

3.3. EXEMPLE DE BOÎTE À OUTIL DES ÉTUDES DE CAS

Chacun a ses préférences pour communiquer sur les études de cas. Ce qui suit est une liste type d'études de cas de référence d'Afrique du Sud. Certains cas de rupture de barrage apparaissent dans plusieurs listes pour souligner des aspects particuliers de la surveillance de barrage (par exemple, Teton, Malpasset, Vajont et plusieurs autres ruptures et / ou incidents). En substance, la boîte à outils type d'Afrique du Sud contient les études de cas suivants :

Barrage de Teton (et barrage de Fontenelle), USA

La rupture du barrage de Teton est suffisamment connue pour être mentionnée au début d'une présentation. Cela « choque » le public, impose le silence et retient toute son attention. C'est aussi un bon exemple pour faire passer le message tout en atténuant les critiques et se mettre en quelque sorte à la place de ses collègues. Un aspect remarquable de la surveillance de ce barrage est que lors du premier remplissage, l'inspection visuelle a même couvert plusieurs centaines de mètres en aval de l'ouvrage.

L'incident du barrage de Fontenelle (qui n'a pas encore été intégré dans les études de cas) s'est produit environ 11 ans avant la rupture du barrage de Teton. Il est parfois ajouté pour certains publics, car une fuite similaire est apparue sur le parement aval. Sur le barrage de Fontenelle, des enrochements (rip-rap) ont été poussés à l'aide d'un bulldozer dans le trou observé. Si des enrochements n'avaient pas été déversés depuis la crête et le réservoir rapidement abaissé, le barrage se serait rompu du fait de la poursuite de l'érosion interne. 20 ans après, une paroi moulée en béton de 61cm d'épaisseur a été insérée dans le noyau du barrage. La paroi a été ancrée dans le rocher de fondation sur 12 à 15 m moyennant l'usage d'une hydro-fraise.

Les leçons tirées de l'incident du barrage de Fontenelle en 1965 n'ont toutefois pas été prises en compte dans les méthodes de conception et de construction de l'USBR. Il semble que l'accident du barrage de Teton soit également imputable à l'érosion des matériaux de remblai à travers les fractures restées ouvertes dans la fondation. L'observation de Ralph Peck relative au manque d'instrumentation au barrage de Teton est à souligner. *« Parmi les reproches avancés après la rupture du barrage de Teton était le manque d'instrumentation. Pourtant, il est extrêmement douteux que des observations instrumentales aient pu alerter en temps utile sur la rupture survenue. Les instruments ne peuvent pas pallier les défauts de conception, ni indiquer de signes de détérioration ou rupture imminentes, sauf s'ils sont fortuitement du type adéquat et installés au bon endroit »* (Dunnicliff & Green 1988, préface).

3.2. AIMS OF BENCHMARK CASE HISTORIES

There are several aims with a set of benchmark case histories. The noble objectives are:

- Primarily, NOT to criticise <u>but to learn from the facts</u> (for example the design of the dam or monitoring system, or the decisions taken by operational staff etc.); and

- Secondarily, <u>to stress points in an unbiased and professional manner</u> and not to point fingers.

In this way, significant universal lessons can be learnt from these case histories and provide guidance on design aspects, construction and operation procedures.

3.3. TYPICAL TOOLBOX OF CASE HISTORIES

Every presenter has his own preferences as far as case histories are concerned. The following is a typical Southern African list of "benchmark" case histories used to bring across certain facts or lessons. Some case histories of dam failures appear on several benchmark lists to stress particular general dam surveillance aspects (for example, Teton, Malpasset, Vajont and several other dam failures and/or incidents). In essence, the typical Southern African "toolbox" contains the following case histories:

Teton Dam (and Fontenelle Dam), USA

Teton Dam failure is a popular case history to start a presentation with, as it "shocks" the audience to silence and gets full attention time and again. It is also a good example to bring the lesson across to temper criticism and rather to put yourself in your colleagues' shoes. A remarkable surveillance aspect is that during first filling, the visual inspection even covered several hundred meters downstream of the dam.

The Fontenelle Dam incident (not incorporated in the case histories yet) occurred about 11 years prior to the Teton Dam failure. It is sometimes added for some audiences, as a similar leak appeared on the downstream face. A bulldozer was used at Fontenelle Dam to push dumped rock (rip-rap) material into the hole at the downstream face. Had rock not been dumped from the crest and the reservoir not been lowered rapidly, the dam would have failed through continuing internal erosion. Two decades later, a 61 cm thick concrete diaphragm wall was added into the core of the dam. It extended between 12 and 15 m below the base of the dam into bedrock, using the "Hydrofraise" system of grout trenching.

The lessons drawn from the 1965 Fontenelle Dam incident were however not incorporated into USBR's design and construction practices. Teton Dam apparently also failed through erosion of embankment materials into unprotected fracturing in the foundation. Ralph Peck's remarks on the lack of instrumentation at Teton Dam is of significance, *"Part of the criticism directed at Teton Dam following its failure was paucity of instrumentation. Yet it is extremely doubtful that any instrumental observations could have given timely warning of the particular failure that occurred. Instruments cannot cure defective designs, nor can they indicate signs of impending deterioration or failure unless, fortuitously, they happen to be of the right type and in the right place"* (Dunnicliff and Green 1988, preface).

Barrage de Malpasset, France (et barrage St Francis, USA)

Le cas du barrage de Malpasset ne met pas l'accent uniquement sur l'importance des investigations géologiques et géotechniques au stade de la conception, mais aussi sur celle de l'implication du concepteur durant toutes les phases du projet, notamment l'évaluation des résultats des mesures d'auscultation. La rupture bien connue du barrage de St Francis (USA), ne figure pas dans la présente liste des études de cas, mais ce fut un cas très parlant en ce qui concerne l'importance, pour les projets de barrages, des investigations géologiques et géotechniques.

Barrage de Vajont, Italie

L'incident du barrage de Vajont est également bien connu pour mériter d'être utilisé en introduction à une conférence ou une présentation en raison des graves conséquences qu'il a occasionnées. Il n'est pas toujours mentionné que des essais sur modèle avaient été effectués pour donner des indications sur la sévérité de l'effet « tsunami » dans le cas où le versant gauche viendrait à glisser dans le réservoir. Cependant, aussi bien l'étendue que la vitesse du glissement ont été largement sous-estimées durant ces essais, ce qui a entraîné une sous-estimation des conséquences. Conformément aux résultats du modèle, le réservoir avait alors été abaissé à la cote « de sécurité » obtenue visant à contenir la vague, mais évidemment, la sous-estimation était considérable. Il est également intéressant de mentionner que la faune du site avait commencé à fuir le site trois jours avant le catastrophique glissement de terrain.

Barrages de Zeuzier et de Nalps, Suisse

Les effets de la pression interstitielle à grandes profondeurs ne sont pas toujours bien compris, comme le montre l'historique documenté du cas du barrage de Zeuzier. Lors de la construction récente du tunnel de base du Saint-Gothard, les taux de réduction de la pression interstitielle et d'infiltration ont été bien surveillés et bien gérés. Des translations et des rotations significatives ont eu lieu sur le site du barrage de Nalps (valeurs mesurées), mais avec peu ou pas de preuves visuelles dans les galeries ou les zones environnantes. Le fait que les exploitants du barrage de Nalps aient effectué et documenté des relevés détaillés des fissures dans les galeries depuis la construction du barrage en 1965 s'est révélé particulièrement intéressant. En effet, leurs graphiques indiquent clairement le développement de nouvelles fines fissures et la progression de fissures existantes.

Barrage de Zoeknog, Afrique du Sud

Le cas du barrage de Zoeknog est remarquable du fait qu'il a mis en exergue l'importance de la compétence et de l'expérience des techniciens chargés de l'installation du dispositif d'auscultation. L'instrumentation (piézomètres) avait prédit de sérieux problèmes durant le premier remplissage. Le mode de rupture potentiel déduit des mesures piézométriques a été ignoré par l'équipe en charge de la surveillance, considérant qu'il s'agissait d'un défaut de fonctionnement. Le barrage s'est rompu quelques semaines après la première alarme déclenchée par le technicien expérimenté en instrumentation. De nombreuses leçons ont été tirées de cette étude de cas, dont certaines peu évidentes. Par exemple, la leçon consistant à ne pas attribuer le contrat concernant l'instrumentation au concurrent le moins cher, mais à celui qui dispose des techniciens d'installation les plus compétents. Dans le même registre, il convient d'ajouter le recours aux équipements d'auscultation les plus fiables. Au barrage de Zoeknog, l'entreprise chargée de l'installation était impliquée dans la conception du système d'auscultation et libre de choisir et fournir les capteurs qui lui semblaient les plus appropriés.

Malpasset Dam, France (and St Francis Dam, USA)

The Malpasset Dam case history not only stresses the importance of geotechnical and geological investigations during the design phase of a dam, but also the importance of keeping at least the designer involved during all phases of the project, especially during the evaluation of monitoring results. The well-known St Francis Dam failure is not covered in the present list of case histories, but it was also a watershed case history in the USA as far as geotechnical investigations at dams are concerned.

Vajont Dam, Italy

The Vajont Dam case history is also a very popular case history to use as introduction to a lecture or presentation due to the grave consequences. It is not always mentioned that model tests had been performed to get an indication of the severity of the "tsunami" effect if the left back would slide into the reservoir. The size and speed of the slide were however significantly underestimated during the model tests. These underestimations contributed to the underestimation of the consequences. The reservoir level was lowered to the calculated "safe" water level in the reservoir to contain the flood wave as predicted by the model tests, but it was obviously grossly under-estimated. It is also interesting to mention that the wild animals started fleeing the area three days prior to the catastrophic landslide.

Zeuzier Dam and Nalps Dam, Switzerland

The effects of deep-seated pore pressures are not always appreciated, as clearly demonstrated by documented Zeuzier Dam case history. During the recent construction of the Gotthard basis tunnel, the pore pressure relief and seepage rates were well monitored and managed. Significant translations and rotations occurred at the dam site of Nalps Dam (measured values), yet with no or little visual evidence in galleries nor the surrounding areas. The fact that the local dam operators of Nalps Dam performed and documented detailed crack surveys in the galleries since construction of the dam in 1965 is noteworthy. Their graphs clearly marked the development of new hairline cracks and progressing of existing cracks.

Zoeknog Dam, South Africa

Zoeknog Dam case history is remarkable in the sense that it stresses the importance of a diligent and experienced instrumentation technician (installer). The instrumentation (piezometers) predicted serious problems during first filling. The potential failure mode indicated by the piezometer was ignored as it was considered as a sensor failure by the site supervision staff. The dam failed within a few weeks after the experienced instrumentation technician's first warning. Numerous lessons are learnt from this case history, some of which are not so obvious. For example, the lesson that instrumentation should not be awarded on the basis of the lowest bidder but awarded to the most diligent installer. In the same vein, the use of the most reliable instruments should be added. At Zoeknog Dam the installer provided his input during the design of the monitoring system and had a free hand to choose and procure the sensors he preferred.

Barrage de Cahora Bassa, Mozambique

Le cas de l'instrumentation du barrage de Cahora Bassa est habituellement présenté pour insister sur l'importance, non seulement pour le long terme d'une conception et d'une installation adéquates, comme cela a été discuté pour le cas du barrage de Zoeknog, mais aussi, pour une équipe de surveillance dédiée, chargée notamment du contrôle et de la maintenance du système d'auscultation ainsi que de sa mise à niveau. Il convient également de mentionner la surveillance vibratoire sous bruit ambiant ainsi que le contrôle du système de gestion des données d'auscultation.

Barrage de Kouga, Afrique du Sud

Le barrage de Kouga est le siège d'une faible réaction alcali-agrégat et est régulièrement cité pour mettre l'accent sur l'importance du suivi par des fissuromètres 3D disposés sur les joints horizontaux de construction (levées de bétonnage) ainsi que la surveillance vibratoire en vue :

- D'identifier les changements brutaux dans le comportement structurel traduisant des désordres éventuels de la structure.

- De mettre l'accent sur l'importance de l'auscultation statique et dynamique en 3D, de sorte à avoir une bonne appréciation des propriétés de la structure et de son comportement statique et dynamique.

- D'insister sur le fait que le suivi des vibrations naturelles apporte un lien pertinent entre la structure en fin de construction et les modèles numériques théoriques (à travers les propriétés dynamiques comme les fréquences naturelles et les modes propres) dans le but de développer un modèle éléments finis étalonné sur les conditions statiques et dynamiques.

Barrage Albasini, Afrique du Sud

L'incident du barrage Albasini est un cas typique de défaillance du système d'alimentation de secours. Le groupe de secours était en parfait état de marche, mais était raccordé à la même ligne que le réseau d'alimentation en énergie des équipements hydromécaniques. Un arbre a endommagé la ligne, si bien que le groupe de secours n'a pas pu servir pour la manœuvre des vannes de l'évacuateur en crête. Heureusement, un électricien résident sur le barrage était au courant de l'existence d'un troisième système de secours opérant avec un moteur diesel. Après remise en service du moteur, non utilisé depuis plusieurs années (dans le noir, sous la lumière des allumettes), le système a pu fonctionner juste à temps pour l'ouverture de la première vanne et éviter ainsi le déversement et la rupture.

Barrages Driekoppies et Inyaka, Afrique du Sud

Le cas du barrage Driekoppies met en exergue la nécessité d'une surveillance permanente de toute perforation dans un remblai ou sa fondation (en cas de forage pneumatique, la pression de l'air au débouché de l'outil de perforation doit être réglée à une valeur égale ou légèrement supérieure à la pression géostatique au droit du fond du trou, de manière à chasser les déblais de forage sans fracturer le sol).

Le cas des deux barrages Driekoppies et Inyaka a montré que les cellules de pression totale peuvent fournir des informations utiles si elles sont convenablement conçues et installées en groupes. Il faudra veiller à pourvoir chaque groupe d'une redondance suffisante pour pouvoir déterminer les contraintes principales en cas de dysfonctionnement d'une, voire deux cellules. L'intérêt d'une présentation judicieuse des résultats des mesures d'auscultation est également mis en évidence.

Cahora Bassa Dam, Mozambique

The case of the instrumentation of Cahora Bassa Dam is usually added to emphasize not only the long-term value of diligent design and installation, as discussed in the Zoeknog Dam case history, but also the value of dedicated dam surveillance staff to carry out observations and perform maintenance as well as regular upgrading of the monitoring system. Their continuous Ambient Vibration Monitoring (AVM) and surveillance data management system are also worth mentioning.

Kouga Dam, South Africa

The Kouga Dam is subjected to mild AAR (Alkali-Aggregate-Reaction) and the case history is regularly used to highlight the value of regular 3-D crack gauge readings across horizontal construction lifts, as well as AVM monitoring in order to:

- Identify sudden changes in structural behaviour and possible structural deterioration.

- Indicate the importance of tri-axial static and dynamic monitoring to get a better picture of the static and dynamic behavior and properties of the structure.

- Stress the fact that AVM can provide the vital link between the as-built structure and the theoretical finite element model (via dynamic properties such as natural frequencies and mode shapes) in order to develop a single FEM (Finite Element Model) that is calibrated for both static and dynamic conditions.

Albasini Dam, South Africa

The incident at Albasini Dam is a typical version of a back-up power system that failed. The standby generator was in mint condition but used the same power lines as the grid to supply power to the radial gates. A tree damaged the power lines, resulting in no backup power to operate the flood gates on the crest. Fortunately, an electrician that had sought shelter at the dam was aware of a third mechanical back-up system using a petrol driven motor. After "servicing" the motor that had been in disuse for several years (practically by match light in the dark), they finally got it running just in time to open the first of the radial gates, and thus in time to prevent an overtopping failure.

Driekoppies Dam and Inyaka Dam, South Africa

The Driekoppies Dam case history clearly stresses the fact that drilling into an embankment or its foundation should preferably be done under "around-the-clock" supervision (if a pneumatic drill is used the air pressure at the tip of the drill should be regulated to a value equal to just more than the weight of the soil above the tip in order to be able to blow the drilled material out of the hole without fracturing the soil).

Both Driekoppies and Inyaka Dam case histories proved that earth pressure cells can provide useful information if they are properly designed and installed in clusters. Care should be taken to provide enough redundancy in each cluster in order to determine principal stresses in case one or more of the earth pressure cells are not functioning correctly. Clever ways to present instrumentation results are also demonstrated.

Barrage Gariep, Afrique du Sud

Le cas du barrage Gariep n'a pas été spécifiquement examiné dans les études de cas. Un des aspects les plus intéressants de son comportement est le déplacement permanent de la voûte vers l'aval après une longue période de basses eaux. Cela a été attribué au retrait du béton maintenu sec. Le même comportement a été observé sur d'autres barrages voûte en Afrique du Sud.

Barrage Sayano-Shushenskaya, Russie

L'accident survenu à l'usine hydroélectrique de Sayano-Shushenskaya n'a pas encore été répertorié dans les études de cas. L'usine a été mise en service officiellement en 2000, avec une puissance installée de 6 400 MW, ce qui la classait, à son achèvement, au premier rang en Russie et au 6ème rang mondial. Le projet comprend un barrage poids arqué de 242 m de hauteur. Un grave accident eut lieu le 17 août 2009, avec inondation de l'usine et perte de 75 vies humaines. Plusieurs leçons en ont été tirées notamment en ce qui concerne l'exploitation du barrage. Les désordres ayant affecté l'usine et l'impossibilité de faire transiter l'eau à l'aval par les turbines ont conduit à une réduction notable de la capacité d'évacuation des crues. Ce type d'accident doit être pris en compte dans l'évaluation de la sécurité du barrage et sa surveillance. Les conditions climatiques sévères ayant régné après l'accident ont mis en évidence l'importance d'un examen détaillé des procédures d'exploitation et de maintenance pour tous les modes de rupture envisagés. La vérification de l'efficacité des procédures de gestion de l'évacuateur de crue durant les périodes de température extrême est à intégrer dans les procédures de surveillance.

Barrage Shih-Kang, Taiwan

Le barrage Shih-Kang situé à Taiwan est un seuil poids équipé de vannes secteur. C'est un cas bien connu, mais qui ne fait pas (encore) partie des études de cas répertoriées. Le barrage de 25 m de hauteur a été achevé en 1977 en vue du contrôle des crues et de l'irrigation. Le séisme de Chi-Chi, survenu le 21 septembre 1999 à Taiwan avait un épicentre à 8 km de profondeur et 150 km au sud de Taipei. La secousse principale était de magnitude 7,3, provoquant des désordres non réparables à environ 10 000 bâtiments, 2 400 décès, 10 000 blessés et 100 000 sans-abris. Des mouvements importants se sont produits le long d'une faille qui court le long de la vallée. Des mouvements de plus de 6 m dans le sens vertical et amont aval ont affecté les deux vannes secteur en rive droite. Le barrage est cependant resté opérationnel comme source d'eau d'irrigation.

Un batardeau en remblai a ensuite été réalisé pour éviter le passage de l'eau dans la « brèche » alors que la section endommagée a été conservée comme mémorial. La capacité du réservoir a été réduite de manière à ne plus être utilisée pour le contrôle des crues, tout en restant une source importante pour l'irrigation.

Gariep Dam, South Africa

Gariep Dam case history was not specifically discussed in the case histories. One of the more interesting aspects in its behavior is the permanent downstream displacement of the arch after prolonged periods of low water levels. This has been attributed to drying shrinkage of the concrete and this behavior is also evident at a number of other arch dams in South Africa.

Sayano-Shushenskaya Dam, Russia

The accident at Sayano-Shushenskaya hydropower plant (HPP) is not yet covered in the case histories. It was officially commissioned in 2000, and with an installed capacity of 6 400 MW it is the largest HPP in Russia and the 6th largest HPP in the world (at the time of completion). It consists of a 242 m high concrete arch-gravity dam officially commissioned in 2000. The fatal accident occurred on 17 August 2009, flooding the power station. A number of lessons can be learnt from this case history, where 75 lives were lost. Operational problems were experienced at Sayano-Shushenskaya Dam, following the 17 August 2009 accident. From a dam surveillance point of view, the long-term outlet capacity of flood outlet works should be evaluated. With the HPP failure and consequent loss in outlet capacity of the HHP, the total outlet capacity of the dam was reduced significantly. Severe climatic conditions that followed the HHP failure stressed the importance of detailed maintenance and operation procedures for all envisaged failure modes. Monitoring of the effectiveness of the spillway operation procedures during extreme temperatures should therefore also be incorporated in the surveillance procedures.

Shih-Kang Dam, Taiwan

The Shih-Kang Dam located in Taiwan is an ogee-shaped gravity section dam with radial crest gates and is also a popular case history not covered in the case histories (yet). The 25 m high dam was completed in 1977 for flood control and irrigation purposes. The Chi-Chi earthquake occurred on 21 September 1999 at a depth of 8 km, with its epicenter 150 km south of Taipei, Taiwan. The magnitude of the main shock was 7.3, resulting in about 10 000 buildings irreparably damaged with a death toll of 2 400; 10 000 injured, and more than 100 000 homeless. Excessive movements occurred along a fault section running in an upstream-downstream direction. Vertical and upstream downstream movements in an excess of 6 m occurred through the first two radial gates at the right flank. Yet, the dam remained operational as a source of agricultural water.

Subsequently, an embankment cofferdam was built to prevent water from flowing through the "breach", while the collapsed section has been retained as a memorial. The storage capacity of the dam has been reduced to such an extent that it is no longer used for flood control but remains an important source of agricultural water.

L'utilité de la bio-localisation comme technique d'investigation géophysique rapide a été mise en évidence à travers le cas du barrage Ohrigstad. Les résultats d'une campagne préliminaire de bio-localisation (alors considérée au même titre que la méthode des sourciers et qui ne prend que quelques minutes) étaient annoncés comme une illustration humoristique, à la fin d'une présentation concernant les résultats d'essais géophysiques menés au barrage Ohrigstad en 1979. Quand vint la présentation lors d'une conférence en 1982, il a été constaté que les résultats de la bio-localisation étaient en accord avec ceux des autres méthodes. Ce qui devait servir pour introduire de l'humour dans la présentation s'est révélé en fait un véritable outil d'auscultation. Ainsi la méthode est utilisée depuis plusieurs décennies en Afrique du Sud, et avec succès pour l'évaluation de la sécurité des barrages.

Il convient de noter que la bio-localisation n'est pas reconnue comme une méthode scientifique, bien qu'elle soit utilisée par de nombreux « scientifiques » (ingénieurs, géologues, etc.). Comme d'autres méthodes similaires, elle était utilisée dans l'Ex Union Soviétique et fait l'objet de publications dont certaines, traduites, portant sur l'exploration minière par la méthode de la bio-localisation.

Ohrigstad Dam, South Africa

The usefulness of bio-location as a fast "geo-physical" type of investigation technique has been demonstrated with the Ohrigstad Dam case history. The results of a preliminary bio-location survey (then called "water divining" and only taking a few minutes) were intended to be used as a humorous ending for a presentation of geophysical test results in 1979 at the Ohrigstad Dam. When time came to use these bio-location results for the presentation at a conference in 1982, it was discovered that the bio-location results in fact tie in well with the results of the other methods. What was supposed to be a humorous ending to the presentation backfired to become a valuable monitoring tool. In South Africa, it has however been used successfully in dam safety evaluations for several decades.

Note that bio-location is not a recognised scientific method, despite the fact that it is acknowledged and used by many "scientists" (engineers, geologists etc.). However, this and similar methods have been used in the former Soviet Union as published in a few translated papers on ore exploration by means of bio-location.

4. REMARQUES FINALES

4.1. GÉNÉRALITÉS

L'auscultation est une chaîne d'activités interdépendantes. Quelle que soit la manière dont le système d'auscultation a été conçu et mis en œuvre, il ne sera d'aucune utilité en l'absence d'une bonne maintenance, de relevés et d'une interprétation régulière des mesures. De même, des inspections visuelles et des analyses des données de surveillance qui ne seraient pas convenablement conduites et documentées, et menées par un personnel consciencieux et expérimenté, peuvent être d'une portée limitée.

La quantité de données peut être d'une grande utilité dans la détermination du comportement à court terme. Cependant, une énorme quantité de données ne doit pas être considérée comme un indicateur d'une bonne surveillance. Des ensembles volumineux de données et des graphiques non interprétés donnent la fausse impression que les bonnes pratiques de la surveillance sont appliquées.

Ce sont plutôt la qualité des données, la qualité de leur traitement et la qualité de leur évaluation et de la diffusion de l'information qui devraient être des indicateurs de la qualité de la surveillance.

4.2. ÉTUDES DE CAS

Le Comité technique de la surveillance des barrages est convaincu d'avoir documenté suffisamment d'études de cas pour tirer les enseignements nécessaires aux principes de base de la surveillance. Le nombre de cas aurait pu être réduit pour éviter des redondances dans les leçons tirées, mais il a été finalement décidé de ne rejeter aucun cas dès lors qu'ils se limitent à une page chacun dans la version imprimée (la version numérique acceptant, à des fins pratiques, jusqu'à 10 pages pour chaque cas d'étude).

La distribution géographique des cas d'étude est incomplète du fait qu'elle ne couvre pas les incidents survenus dans de grands pays constructeurs de barrages comme la Chine, la Russie, le Brésil ou la Turquie. Certaines études de cas mettent en évidence le comportement du barrage. Les apports positifs des systèmes d'auscultation sont très largement présents, alors que les défauts d'instrumentation ou les problèmes liés aux systèmes de surveillance sont minoritaires.

De nombreux enseignements sont apparus à la préparation du rapport, comme par exemple :

- La technologie numérique s'est développée et est devenue disponible récemment, entraînant des modifications des capteurs et des équipements d'acquisition de nouvelle génération ;

- Les équipements numériques ont une durée de vie relativement courte avant que leurs composants électroniques ne deviennent obsolètes, nécessitant une mise à niveau plus fréquente que la génération précédente d'instruments ;

- On observe actuellement une tendance à accumuler plus de données, mais comparativement moins d'informations et d'évaluations.

4. CONCLUDING REMARKS

4.1. GENERAL REMARK

Monitoring is an interdependent chain of activities. No matter how well an instrumentation system has been designed and installed, their purpose will be nullified if they are not well maintained, observed and the results evaluated regularly. Similarly, visual inspections and surveillance related analyses not performed and properly documented by diligent and experienced persons may be of limited value.

Quantity of data can be very useful to determine short-term behaviour, but huge quantities of data should not be mistaken as a measure of the quality of surveillance. Massive sets of data and un-interpreted line graphs may provide a false sense of confidence that good surveillance practices are being followed.

Quality data, quality data processing, and quality evaluation and dissemination of information should rather be the measure of the quality of the dam surveillance practice.

4.2. CASE HISTORIES

The TCDS is confident that we have documented enough case histories to learn the basic surveillance lessons. The number of case histories documented could have been reduced to avoid repetition in lessons, however, it was finally decided not to reject any case history as they are limited to the one page in the printed version. On the other hand, the digital version of the bulletin has for practical purposes a 10 page limitation for case histories.

The geographical spread of the case histories is incomplete (despite several attempts and promises) as it does do not cover incidents in major dam building countries such as China, Russia, Brazil and Turkey. Some of the case histories tend to highlight behaviour of the dam. Successful surveillance systems dominate the case histories, whereas instrumentation failures or problems with the surveillance systems are in the minority.

Several lessons surfaced during the preparation of the bulletin, for example:

- Significant digital technology has become available recently resulting in changes in several of the new generation sensors and read-out equipment

- Digital equipment has a relative short life span before their electronic components becomes obsolete, requiring more frequent upgrading than the previous generation of instruments

- There is currently a tendency to accumulate more data but relatively less information and evaluation

DOI: 10.1201/9781003274841-4

4.3. LES SUITES À DONNER

Il est prévu que :

- Une version imprimée condensée et une version numérique complète du bulletin soient disponibles sur le site internet de la CIGB (une version imprimée abrégée au format «PDF» et une version numérique plus volumineuse comprenant les versions complètes des études de cas, ainsi qu'un tableau Excel qui en donne la classification. Ce dernier est destiné aux utilisateurs qui souhaitent effectuer leurs propres sélections) ;

- Que le présent Comité technique de la surveillance des barrages, parallèlement à la préparation du deuxième bulletin intitulé « Acquisition et interprétation des données et des observations de la surveillance », ait la possibilité de répertorier davantage d'études de cas pour enrichir ce bulletin (et d'en publier les mises à jour significatives à l'avenir).

4.4. UN TRAIT D'HUMOUR POUR TERMINER

La version imprimée de ce bulletin est plutôt courte et la version numérique longue, car si nous avions eu plus de temps, nous aurions écrit une version écrite plus longue et une version numérique plus courte. (Nos excuses à Blaise Pascal qui a écrit dans « Lettres Provinciales » en 1657 : « *Je n'ai fait celle-ci plus longue que parce que je n'ai pas eu le loisir de la faire plus courte* »).

4.3.   THE WAY FORWARD

It is suggested that:

- Both a short printed and long digital version of the bulletin be made available on ICOLD web site (short printed version in "pdf" format and the much "larger" digital version that includes the full versions of the case histories, as well as an excel spreadsheet file showing the classification of the case histories. The latter is for users who want to make their own selections; and

- That the present TCDS, parallel to preparing the second bulletin, allow for more case histories for this bulletin (and publish significant updates as revisions in future).

4.4.   TONGUE IN THE CHEEK FINAL REMARK

The printed version of this bulletin is quite short and the digital version so long, because if we had had more time, we would have written a longer printed version and a shorter digital version. (Apologies to Blaise Pascal who said in the year 1657: *"Je n'ai fait celle-ci plus longue que parce que je n'ai pas eu le loisir de la faire plus courte"* in "Lettres Provinciales").

5.   RÉFÉRENCES

E. DIBIAGIO (2013) Field Instrumentation - The Link between Theory and Practice in Geotechnical Engineering. Conference proceedings. Seventh (7th) International Conference on Case Histories in Geotechnical Engineering: and Symposium in Honor of Clyde Baker, April 29 - May 4, 2013. Wheeling (Chicago)

J. DUNNICLIFF with the assistance of G. E. GREEN (1988) Geotechnical instrumentation for monitoring field performance, Wiley-Interscience Publication

ICOLD (1988). *Dam monitoring, General considerations*. Bulletin 60. Paris, France.

ICOLD (1989). Monitoring of dams and their foundations, State of the Art. Bulletin 68. Paris, France.

ICOLD (1992). *Improvement of existing Dam Monitoring*. Bulletin 87. Paris, France.

ICOLD (1995). *Dam failures – Statistical Analysis*. Bulletin 99. Paris, France.

ICOLD (2000). Automated Dam Monitoring Systems, Guidelines and case histories. Bulletin 118. Paris, France.

ICOLD (2009). *General Approach to Dam Surveillance*. Bulletin 138. Paris, France.

ICOLD (2013 English and 2016 French). *Dam Surveillance Guide*. Bulletin 158. Paris, France.

SPANCOLD (2012) *Risk Analysis applied to Management of Dam Safety*, Technical Guide on Operation of Dams and Reservoirs Vol. 1.

D'autres références associées à chaque étude de cas sont identifiées dans la version numérique.

5.   REFERENCES

E. DIBIAGIO (2013) Field Instrumentation - The Link between Theory and Practice in Geotechnical Engineering. Conference proceedings. Seventh (7th) International Conference on Case Histories in Geotechnical Engineering: and Symposium in Honor of Clyde Baker, April 29 - May 4, 2013. Wheeling (Chicago)

J. DUNNICLIFF with the assistance of G. E. GREEN (1988) Geotechnical instrumentation for monitoring field performance, Wiley-Interscience Publication

ICOLD (1988). *Dam monitoring, General considerations*. Bulletin 60. Paris, France.

ICOLD (1989). Monitoring of dams and their foundations, State of the Art. Bulletin 68. Paris, France.

ICOLD (1992). *Improvement of existing Dam Monitoring*. Bulletin 87. Paris, France.

ICOLD (1995). *Dam failures – Statistical Analysis*. Bulletin 99. Paris, France.

ICOLD (2000). Automated Dam Monitoring Systems, Guidelines and case histories. Bulletin 118. Paris, France.

ICOLD (2009). *General Approach to Dam Surveillance*. Bulletin 138. Paris, France.

ICOLD (2013 English and 2016 French). *Dam Surveillance Guide*. Bulletin 158. Paris, France.

SPANCOLD (2012) *Risk Analysis applied to Management of Dam Safety*, Technical Guide on Operation of Dams and Reservoirs Vol. 1.

Further references are given in each case history included in the digital appendix.

DOI: 10.1201/9781003274841-5

6.1. LIST OF CASE HISTORIES WITH MAIN OBJECTIVE, MAIN BENEFIT AND OBSERVATION

N°	Country	Dam name	Dam type	Description	Main Objective	Main Benefit	Observations
Benchmark case histories							
1	France	Malpasset	Arch	Double curvature concrete	Importance of monitoring engineering geological aspects	Development of rock mechanics applied to dam engineering	A milestone case history highlighting importance of dam surveillance
2	Italy	Vajont	Arch	Double curvature concrete	Importance of monitoring reservoir slopes	Improvement of field investigations and slope stability analysis	Emergency action plans have become mandatory in many countries to mitigate consequences of dam failure
3	Switzerland	Zeuzier	Arch	Double curvature concrete	The unbelievable effect of pore pressure relief	Correct remedial action preventing serious incidents	Importance of diligent surveillance, data analysis and evaluation
4	USA	Teton	Embankment	Zoned earthfill with central clay core	The value of diligent visual observations	The need for site specific dam design became obvious	Importance of failure mode-oriented surveillance and emergency action plans
5	USSR / Germany	Dnieprostri / Möhne / Eder	Gravity / gravity / gravity	Concrete / masonry / masonry	The impact of explosive loads during World War II	Geneva convention states that dams and dykes should not be object of military attack	Dams can be damaged not only due to natural conditions but also by man created forces

(Continued)

DOI: 10.1201/9781003274841-6

Nº	Country	Dam name	Dam type	Description	Main Objective	Main Benefit	Observations
6	USA	Folsom	Gravity & embankment	Concrete & zoned earthfill with central silty core	Gate failure although tested regularly but not all the way	Better understanding of structural behaviour of spillway gates	Various re-evaluation programs of similar gates were carried out after the Folsom incident
7	Mozambique	Cahora Bassa	Arch	Double curvature concrete	The value of diligent installations on the life of instruments	Best monitoring practice to identify and assess incidents (AAR)	Continuous dynamic monitoring (AVM) is considered very appropriate for arch dams
8	South Africa	Zoeknog	Embankment	Zoned earthfill with central clay core	Failure predicted by pore pressure gauges, but ignored	Importance of diligent monitoring data assessment	Value of appropriate and experienced monitoring staff
9	Spain	Tous	Embankment	Rockfill with clay core	Highlights the backup systems failure	Importance of reliable and redundant energy sources	Tous dam break lead to a complete review of Spanish dam safety regulations

Other case histories

Nº	Country	Dam name	Dam type	Description	Main Objective	Main Benefit	Observations
1	Argentina	El Chocón	Embankment	Zoned earthfill with central clay core	Early detection of failure mechanisms	Correct and on time remedial action	Early detection of potential erosion and assessment of remedial works
2	Austria	Durlassboden	Embankment	Zoned earthfill with central clay core	Surveillance of the grout curtain	Good understanding of dam behaviour	Control of seepage and piezometric levels in foundation
3	Austria	Gmuend	Gravity	Concrete	Influence of material of instruments on results	Improvement of measurement results	Temperature impact on quality of monitoring data
4	Austria	Zillergründl	Arch	Double curvature concrete	Surveillance of dam foundation & seismic analysis	Controlled uplift in the dam foundation & confirmation of numerical analysis	Good correlation of predicted and measured dam behaviour
5	Burkina Faso	Comoé	Embankment	Homogeneous earthfill	Highlight the specific behaviour of tropical soils	Improving the safety of the dam without emptying the reservoir	Detection of piping action in a lateritic dam foundation due to erosion and dissolution

(Continued)

117

6	Cameroun	Song Loulou	Embankment with gravity spillway	Earthfill & concrete	Close follow-up of concrete expansion due to AAR development	After its reinforcement, monitoring system became efficient	AAR is a very serious concrete pathology which may impact dam safety and operation
7	Canada	Kootenay Canal Forebay	2 embankments	Both concrete faced rockfill	Leakage analysis at joints (concrete slabs and plynth)	Characterization of leakage and design of remedial works	Importance of persistent, long-term investigations and weir flow measurements by surveillance personnel
8	Canada	WAC Bennett (trends analysis)	Embankment	Zoned earthfill with central clay core	Regression analysis weir flow and reservoir levels	Performance evaluation of instruments and prediction	Regression analysis considering lag time and creep leads to more precise results, long term weir flow measurements
9	Canada	WAC Bennett (ROV)	Embankment	Zoned earthfill with central clay core	Identify and monitor features in the underwater portion of the dam	Create a full 3-dimensional model of the upstream face of the dam	Simultaneous multi-beam and side scan baseline surveys
10	Canada	Not reported	Embankment	Sand/gravel fill with central till core	Geostatistical analysis to assess a zone of higher hydraulic conductivity in a central till core	Decide if remedial actions were required	Importance of construction control data and as-built reports for analysis of monitoring data
11	Canada	Not reported	Embankment	Zoned earthfill with central clay core	Minimize internal erosion in an underlaying sand layer	Design of targeted relief wells as stabilizing measure	Electromagnetic and Lidar surveys as well as geostatistical analysis
12	Colombia	Porce II (assessment study)	Gravity & embankment	RCC & homogeneous earthfill	Evaluate the safety of the dam	Detect and timely attend deficiencies in the dam	Analysis of monitoring data allowed to evaluate the dam safety conditions

(*Continued*)

13	Colombia	Porce II (thresholds)	Gravity & embankment	RCC & homogeneous earthfill	Definition of monitoring thresholds	Detect and prevent the development of possible failure modes (FM)	The definition of alert or alarm thresholds is fundamental to identify FM that might affect the safety of the dam
14	Colombia	Santa Rita	3 embankments	Earthfill – silty core at main dam	Early detection of failure mechanisms	Evaluate the effectiveness of the rehabilitation or upgrade works	Reduction of liquefaction potential and assessment of upgrade works
15	Colombia	Tona	Embankment	Concrete faced rockfill	Comparison of the deformation modulus of laboratory and theoretical values with field data	Laboratory tests could reproduce well the rockfill deformation modulus	A good approximation of the modulus allows one to optimize the design and lower the risks of the design
16	Colombia	Urrá I	Embankment	Earthfill with clay core	Early detection of failure mechanisms	Correct and on time remedial action	Early detection of landslide and assessment of remedial works
17	Czech Republic	Mšeno	Gravity	Masonry	Improvement of dam safety due to ageing of dam foundation of a 100-year-old dam	Bedrock sealing made from new grouting gallery	Long-term development of seepage regime
18	Egypt	El-Karm	Gravity	Concrete	Early detection of crack mechanisms	Correct and on time remedial action	Differential settlements cause cracks due to missing construction joints, assessment of remedial works
19	France	Etang	Embankment	Zoned earthfill and rockfill with upstream PVC lining	Importance of visual inspection in association with monitoring	Appropriate and early actions engaged for dam safety	Interest of alarms for early detection of events concerning dam safety

(Continued)

20	France	Grand'Maison	Embankment	Rockfill with central clay core	Reservoir landslide monitoring and risks in dam environment	Safe operation of the dam with the slide above the reservoir	Evolution of measurements (geodesy, photogrammetry, GNSS, InSar)
21	France	La Paliere	Embankment	Homogeneous earthfill	Maintenance linked to monitoring	Keeping dam safety at a controlled level	Visual inspections are of primary importance for dikes as well as the assessment of slowly evolving phenomena
22	France	Mirgenbach	Embankment	Homogeneous earthfill	Monitoring water pressure during construction	Control of water pressure dissipation and stability of embankments	Sliding rupture of both embankments not detected on time (badly installed pore pressure meter)
23	Germany	Sylvenstein	Embankment	Zoned earthfill with central clay core	Improvement of the sealing system and the seepage flow measurement	Improvement of the dam safety standard	Innovative methods (deep cut-off wall, additional control gallery)
24	Iran	Alborz	Embankment	Rockfill with central clay core	Replacement of vibrating wire piezometers instead of standpipes	Longer instrument durability for dam body with high deformation	Around 60% of standpipes have been distorted due to deformation of the dam body mostly during construction
25	Iran	Gotvand	Embankment	Zoned rockfill with central clay core	Verify design and behaviour follow-up during construction, impoundment and operation	Adding supplementary instrumentation sections during construction	Excess pore water pressure in clay core and the maximum pore water pressure coefficient at the largest dam section
26	Iran	Karun 4	Arch	Double curvature concrete	Early reorganization of dam body cracks	Rehabilitation of some of the cracks	The trend of cracks opening, and rehabilitation (reinjection) performance (efficiency) has been controlled

(Continued)

No.	Country	Dam name	Dam type	Type			
27	Iran	Masjed-e-Soleiman	Embankment	Rockfill with central clay core	Controlling dam transient condition	Indicating that the dam behaves consistent but not satisfactory	Continuous behaviour with abrupt or sudden changes
28	Iran	Seymareh	Arch	Double curvature concrete	Response of the dam instruments to internal and external factors	Ensuring of the dam instrumentation system	PMF-orientated and very well performing system (99,7% of the devices working) - main load is thermal
29	Italy	Ambiesta	Arch	Double curvature concrete	Detection of cracking patterns inside the dam body	Early detection of cracking patterns variations or extension	Investigations with geophysical techniques such as sonic tomography and micro seismic refraction
30	Italy	Isola Serafini	Buttress/gravity weir	Concrete	Early detection of failure mechanisms: erosion assessment of downstream riverbed	Correct and on time remedial action	Monitoring of erosion of the downstream riverbed by multibeam bathymetry technique
31	Italy	San Giacomo	Buttress	Concrete	Stability analysis	Uplift force to verify sliding condition of the structure	Control of slide failure mode
32	Italy	Several dams (MISTRAL)	All dam types	-	Innovate data processing and presentation techniques	Support surveillance tasks and decision-making for dam safety management	Decision Support System based on automatic monitoring
33	Japan	Kyogoku Upper Reservoir	Embankment	Asphalt faced rockfill	Early detection of leakage from the asphalt facing	Early and quick detection of the location of leakage	Significant reduction of repair costs and repair work period
34	Japan	Okuniikappu	Arch	Double curvature concrete	Faster delivery of higher-quality information	Legal compliance, labor saving, safety enhancement	Automated topographic survey for a remote dam - Immediate dam safety assessment after earthquake

(Continued)

					Identification of crack initiation mechanism in asphalt facing	Early confirmation of the stability of the dam body and early reinforcement	Visual confirmation of damage and reinforcement work of the asphalt facing in a short period of time
35	Japan	Yashio	Embankment	Asphalt faced rockfill	Identification of crack initiation mechanism in asphalt facing	Early confirmation of the stability of the dam body and early reinforcement	Visual confirmation of damage and reinforcement work of the asphalt facing in a short period of time
36	Morocco	Oued El Makhazine (cutoff wall)	Embankment	Rockfill with central clay core	Early detection of failure mechanisms	Correct and on time remedial action	Pore pressure evolution in foundation - early detection of potential erosion and assessment of remedial works
37	Morocco	Oued El Makhazine (culvert joint)	Embankment	Rockfill with central clay core	Systematic visual inspections of all works	Correct and on time remedial action	Piping through embankment - water stop defects in a culvert crossing a clay core should be considered
38	Morocco	Tuizgui Ramz	Gravity	Masonry	Insufficient design quality and surveillance are detrimental for dam safety	Strong awareness of the quality of design, even for dams with limited height	Expert geologists should be mobilised to check the quality of the foundation, despite the type and height of the dam
39	Norway	Muravatn	Embankment	Rockfill with central moraine till core	Evaluate the nature and extent of problem	Correct remedial action was taken	Early detection of potential erosion and assessment of remedial works
40	Norway	Storvatn	Embankment	Rockfill with central inclined asphaltic concrete core	Verify new design concept	Design verified	Construction control of asphalt concrete cores
41	Norway	Svartevann	Embankment	Rockfill with central moraine till core	Appraisal of the stability and performance of a zoned dam	Construction control, long-term monitoring and validation of the design concept	Pore pressures build up and dissipation, leakage, relative movements of different materials or zones within the dam

(Continued)

No.	Country	Name	Type				
42	Norway	Trial dam	Embankment	Rockfill with central bitumen membrane	Verify new design for impervious core	Design verified	Design verification during construction of a new type of membrane
43	Norway	Viddalsvatn	Embankment	Rockfill with central glacial moraine core	Serious turbid leaks, erosion of the core in the general vicinity of the two sinkholes	Knowledge about the filters	Early detection of serious turbid leaks
44	Norway	Zelazny Most (Poland)	Tailings	Ring shaped	Input for the observational method	Design modified on basis of measurements	Observational method for failure mode detection
45	Portugal	Automatic monitoring system	Concrete dams	-	Real-time structural safety control	Detect possible malfunctions as early as possible	
46	Romania	Gura Râului	Buttress	Concrete			
47	Romania	Paltinu	Arch	Double curvature concrete	Early detection of failure mechanisms	Correct and on time remedial action	Early detection of failure mechanisms and evaluation of remedial works
48	Romania	Pecineagu	Embankment	Concrete faced zoned rockfill	Early detection of failure mechanisms	Correct and on time remedial action	Early detection of failure mechanisms and evaluation of remedial works
49	Romania	Poiana Uzului	Buttress	Round headed concrete	Detailed analysis of uplift and temperature loads	Definition of annual reservoir variation for safe operation	Early detection of potential displacements and increased drainage phenomenon
50	South Africa	Belfort	Multiple buttressed arch & embankment	Concrete & homogeneous earthfill	Early detection of failure mechanisms, control of remedial works	Correct remedial action	Detection of cracks and ground deformation

(Continued)

51	South Africa	Driekoppies	Embankment & gravity spillway	Zoned earthfill with central clay core & concrete	Early detection of failure mechanisms	Correct remedial action, control of construction and remedial works	Hydraulic and pneumatic fracture, drilling in embankment, pressure cells
52	South Africa	Inyaka	Embankment & trough spillway	Zoned earthfill with central clay core & concrete	Control of contact between embankment and concrete structure	Obtaining pressure distribution and main stresses	Design and installation of pressure cells, graphical representation of data
53	South Africa	Katse	Arch	Double curvature concrete	Early detection of failure mechanisms	Detection of deformations and cracks	Correct installation of monitoring devices by experienced field technicians is fundamental for an adequate performance
54	South Africa	Kouga	Arch	Double curvature concrete	Detection of swelling and structural deterioration	Understanding the influence of dynamic behaviour on structural deterioration	Analysis with dynamic AVM monitoring and detection of AAR
55	South Africa	Ohrigstad	Embankment	Concrete faced rockfill	Leakage detection	Detection of flow paths	Results from bio-location and other techniques matched
56	Spain	La Aceña	Arch gravity	Concrete	Validation of the DGPS system	Useful in dam monitoring and safety programs, valuable complement to other monitoring methods and interesting for dams with difficult access	Detects absolute deformations and millimeter accuracy. Affordable cost

(Continued)

No.	Country	Dam	Dam type	Material/core	Early detection of failure mechanisms	Risk reduction measures and on time remedial actions	Early detection of potential erosion and assessment of remedial works
57	Spain	Caspe II	Embankment	Zoned earthfill with central clay core	Early detection of failure mechanisms - control of gypsum dissolution in foundation	Risk reduction measures and on time remedial actions	Early detection of potential erosion and assessment of remedial works
58	Spain	Cortes	Arch gravity	Concrete	Reservoir slide monitoring in real time	Early warnings in case of detection of movements	Monitoring of slopes around reservoirs is essential for detecting problems
59	Spain	La Minilla + El Gergal	Gravity & arch gravity	Concrete	Upgrading of monitoring system	Cost and time reduction	Partial automation of readings of monitoring devices
60	Spain	La Loteta	Embankment	Zoned earthfill with central clay core & upstream clay blanket and cut-off wall	Early detection of failure mechanisms during first filling	Detection of a failure mode, follow-up and on-time remedial action	Early detection of seepage avoiding potential erosion and construction of an impervious cut-off at left abutment
61	Spain	Siurana	Gravity	Concrete	Improving dam safety allocating minimum investment	Remedial and efficient works with high-cost reduction	Control and monitoring of uplift pressures during remedial works
62	Spain	Val	Gravity	RCC	Early detection of failure mechanisms	Correct and on time remedial action	Long-term control of displacements, seepage, uplift and thermal state of an RCC dam
63	Spain	55 dams in Ebro river basin	All dam types	-	Centralization of monitoring data	Optimization of monitoring data management	Integration of monitoring in the dam safety management program
64	Sweden	Storfinnforsen	Embankment	Zoned fill with central glacial till core and timber sheet pile	Early detection of potential failure mechanisms	Correct and on time remedial action	Early detection of potential timber deterioration and assessment of remedial works

(Continued)

					Information on dam surveillance status	PFMA, definition of monitoring and surveillance needs	
65	Sweden	20 high risk dams	Several dam types	-			Efficient and PFM-orientated monitoring systems
66	Tunesia	Ziatine	Embankment	Zoned earthfill with central clay core	Importance of sealing mats	Temporary remedial action	Early detection of leakage and assessment of remedial works
67	USA	Dorris	Embankment	Homogeneous earthfill	Potential failure mode analysis	Optimization of monitoring actions and risk reduction	A well-designed monitoring program can achieve substantial risk reduction benefits at low costs.
68	USA	Ochoco	Embankment	Homogeneous earthfill	Early detection of failure modes by visual inspection and monitoring	Sudden Piezometer Rise offers alert	Piping through foundation (landslide material)
69	USA	Steinaker	Embankment	Zoned earthfill with central clay core	Control of landslide movements	Monitoring upgrade to follow-up failure mode development	New monuments and piezometers to monitor slope movements when the reservoir is drawn down
70	USA	Wanapum	Gravity & embankment	Concrete & zoned earthfill	Early detection of failure mechanisms	Design of remedial works and upgrade of surveillance	Excessive tensile stresses caused a long, continuous, horizontal crack and significant displacements of spillway structure
71	USA	Wolf Creek	Gravity & embankment	Concrete & earthfill	Statistical analysis of piezometer data	Verification of adequacy of rehabilitation works (cut-off wall)	Multiple linear regression analysis of piezometer data indicates that cut-off wall was successful

6.2. LIST OF CASE HISTORIES WITH HAZARD OR POTENTIAL FAILURE MODE, KEY WORDS AND AUTHORS

N°	Country	Dam	Dam type	Description	Hazard or PFM	Key words	Authors
Benchmark case histories							
1	France	Malpasset	Arch	Double curvature concrete	Foundation rock deformation, uplift	geotechnical investigations, foundation, pore pressure, failure	B. Goguel
2	Italy	Vajont	Arch	Double curvature concrete	Reservoir slope failure	slope stability, landslide, failure	R. Gómez López de Munain & J. Fleitz
3	Switzerland	Zeuzier	Arch	Double curvature concrete	Pore water pressures	Foundation, deformations, settlement, joints, cracks, monitoring system	H. Pougatsch & L. Mouvet
4	USA	Teton	Embankment	Zoned earthfill with central clay core	Seepage, internal erosion in foundation and contact	Hydraulic gradient, seepage, erosion, dam failure	J. Fleitz
5	USSR / Germany	"	Dnieprostri / Möhne / Eder	Gravity / gravity / gravity	Sabotage	Dam breach, intentional dam demolition, dam busters, Geneva Convention	P. Choquet
6	USA	Folsom	Gravity & embankment	Concrete & zoned earthfill with central silty core	Proper operation and maintenance	Tainter gate, spillway gate failure, corrosion, trunnion coefficient of friction	P. Choquet
7	Mozambique	Cahora Bassa	Arch	Double curvature concrete	Aging of concrete, proper installation, and long-term operation	monitoring system, alkali-aggregate reaction (AAR), concrete swelling	I. Tembe & C. Oosthuizen
8	South Africa	Zoeknog	Embankment	Zoned earthfill with central clay core	Internal erosion, hydraulic fracture, pore water pressure	piezometer, piping, dam failure	C. Oosthuizen
9	Spain	Tous	Embankment	Rockfill with clay core	Failure of gated system due to lack of energy backup system	flood, overtopping, spillway gates, redundant energy source, dam failure	J. Fleitz

(Continued)

DOI: 10.1201/9781003274841-7

127

Other case histories

#	Country	Dam	Type	Detail	Issue	Description	Authors
1	Argentina	El Chocón	Embankment	Zoned earthfill with central clay core	Seepage, internal erosion in foundation and contact	internal erosion, rehabilitation, control method, monitoring	F Restelli & A Pujol
2	Austria	Durlassboden	Embankment	Zoned earthfill with central clay core	Seepage, internal erosion in foundation	dam operation, foundation treatment, grout curtain, monitoring, piezometer, pore pressure	Florian Landstorfer & Pius Obernhuber
3	Austria	Gmuend	Arch dam upgraded with a gravity dam		Downstream slope stability, improvement of monitoring systems	extensometer, monitoring, temperature	Florian Landstorfer & Pius Obernhuber
4	Austria	Zillergründl	Arch	Double curvature concrete	Uplift, earthquake	automated monitoring, cracking, finite elements method, mathematical model, monitoring, numerical model, piezometer, pore pressure, seismic resistance	Florian Landstorfer & Pius Obernhuber
5	Burkina Faso	Comoé	Embankment	Homogeneous earthfill	Seepage, internal erosion in foundation and abutments	weathered rock, seepage, seepage paths, underground seepage,	A.F. Chraibi, Morocco & A. Nombré
6	Cameroun	Song Loulou	Embankment with gravity spillway	Earthfill & concrete	Ageing of concrete (AAR and sulphate attack)	concrete swelling, alkali-aggregate reaction (AAR), sulphate attack	A.F. Chraibi

(Continued)

#	Country	Dam	Type	Fill	Problem	Keywords	Authors
7	Canada	Kootenay Canal Forebay	2 embankments	Both concrete faced rockfill	Seepage through concrete slab joints	slab-plinth joint, PVC waterstop, concrete joint leakage, leak detection, atypical leakage event, geomembrane liner, weir flow, differential settlement, energy balance orifice flow equation, seepage flow-reservoir level correlation	Chris Daniel, Peter Gaffran, Mitchell Illerbrun, Mark Aseleson and Tom Stewart
8	Canada	WAC Bennett (trends analysis)	Embankment	Zoned earthfill with central clay core	Seepage through dam and foundation	Instrument performance evaluation, Trend analysis, Regression analysis, Seepage weir flow, Seepage flow monitoring, Instrument reading lag time, Instrument reading creep, Instrument reading performance bound, Key driver variable, Reservoir level correlation, Sinkhole remediation, Compaction grouting	M. Li and D. Siu, BC Hydro
9	Canada	WAC Bennett (ROV)	Embankment	Zoned earthfill with central clay core	Sedimentation	underwater survey, underwater inspection, remotely operated underwater vehicles (ROV), multibeam sonar, side scan sonar, high resolution imagery, Geographic Information System (GIS) software	Gordon Anderlini and Derek Wilson
10	Canada	Not reported	Embankment	Sand/gravel fill with central till core	Seepage, internal erosion in dam body	dam core heterogeneities, annual thermal response, hydraulic conductivity, advection, till core, temperature, fine content, internal erosion, construction control data, geostatistics,	M. Smith, Hydro-Quebec

(Continued)

#	Country	Name	Type	Material	Issue	Keywords	Authors
11	Canada	Not reported	Embankment	Zoned earthfill with central clay core	Uplift, internal erosion in foundation	seepage detection, electromagnetic survey, lidar survey, internal erosion, foundation sand layer, foundation clay layer, pore pressure, uplift pressure, relief wells, geostatistics	M. Smith, Hydro-Quebec
12	Colombia	Porce II (assessment study)	Gravity & embankment	RCC & homogeneous earthfill	Uplift, seepage through dam and foundation	roller compacted concrete, monitoring, safety, failure, performance, settlement, infiltration, uplift	C. A. Cardona, A. A. Duque
13	Colombia	Porce II (thresholds)	Gravity & embankment	RCC & homogeneous earthfill	Uplift, internal erosion in dam foundation	monitoring, thresholds, safety, failure, performance, inspection, dam	A. A. Duque
14	Colombia	Santa Rita	3 embankments	Earthfill – silty core at main dam	Seepage, internal erosion through dam	monitoring, pore pressure, piezometer, liquefaction, rehabilitation, safety, abutment, drainage	D. E. Toscano, C. A. Cordona
15	Colombia	Tona	Embankment	Concrete faced rockfill	Differential settlement	deformation, modulus, instrumentation, large scale test	M. C. Sierra, L. F. Cárdenas
16	Colombia	Urrá I	Embankment	Earthfill with clay core	Downstream dam slope sliding	fill, abutment, erosion, disposal, saturation, piezometer, deformation, landslide	R. J. Piedrahita de León
17	Czech Republic	Mšeno	Gravity	Masonry	Internal erosion in dam foundation	masonry dam, foundation, seepage, grouting gallery	P. Křivka
18	Egypt	El-Karm	Gravity	Concrete	Differential settlement, cracking of concrete	cracks, differential settlement, basin, heat of hydration, joints	North Sinai
19	France	Etang	Embankment	Zoned earthfill and rockfill with upstream PVC lining	Crack in sealing membrane, seepage	monitoring, safety, inspection, leakage, upstream facing	T. Guilloteau

(Continued)

20	France	Grand'Maison Reservoir	Embankment	Rockfill with central clay core	Reservoir slope sliding	monitoring, safety, reservoir slope, landslide	P. Scharff, R. Boudon and P. Rebut
21	France	La Paliere	Embankment	Homogeneous earthfill	Seepage through dam	monitoring, safety, canal, ageing, settlement, heightening	L. Duchesne
22	France	Mirgenbach	Embankment	Homogeneous earthfill	Upstream dam slope sliding	monitoring, safety, construction, gauge, slope stability	T. Guilloteau
23	Germany	Sylvenstein	Embankment	Zoned earthfill with central clay core	Seepage, internal erosion in dam and foundation	glacial, climate change, cut-off wall, control gallery, monitoring	T. Lang, G. Overhoff
24	Iran	Alborz	Embankment	Rockfill with central clay core	Settlements, hydraulic fracture, pore water pressures	standpipe destruction, more thoroughly installation	A. Noorzad
25	Iran	Gotvand	Embankment	Zoned rockfill with central clay core	Settlements, hydraulic fracture, pore water pressures	earthfill dam, first impounding, supplementary instrumentation sections	A. Noorzad
26	Iran	Karun 4	Arch	Double curvature concrete	Concrete cracks	rehabilitation, crack	A. Noorzad
27	Iran	Masjed-e-Soleiman	Embankment	Rockfill with central clay core	Settlements, deformation, pore water pressures	transient conditions, deformations, cracking	A. Noorzad
28	Iran	Seymareh	Arch	Double curvature concrete	Temperature load	thermal loading, pendulums, jointmeters	A. Noorzad
29	Italy	Ambiesta	Arch	Double curvature concrete	Ageing of concrete, cracking	cracks, sonic tomography, concrete	D. Donnaruma, D. Milani, A. Masera

(Continued)

No.	Country	Dam name	Buttress/gravity weir	Concrete	Downstream erosion	bathymetry, riverbed, erosion	P. Gigli
30	Italy	Isola Serafini	Buttress/gravity weir	Concrete	Downstream erosion	bathymetry, riverbed, erosion	P. Gigli
31	Italy	San Giacomo	Buttress	Concrete	Uplift	piezometer measurements, monitoring	P. Valgoi & A. Masera
32	Italy	several dams (MISTRAL)	All dam types	–	Diligent monitoring data analysis	monitoring, safety, data base, threshold	A. Masera
33	Japan	Kyogoku Upper Reservoir	Embankment	Asphalt faced rockfill	Crack in sealing membrane, seepage	measurement system, leakage, repair work	Jun Takano
34	Japan	Okuniikappu	Arch	Double curvature concrete	Earthquake	displacement, automated monitoring	Jun Takano
35	Japan	Yashio	Embankment	Asphalt faced rockfill	Earthquake, cracks in sealing membrane	monitoring, cracking, reinforcement work, earthquake	Jun Takano
36	Morocco	Oued El Makhazine (cutoff wall)	Embankment	Rockfill with central clay core	Pore water pressures in foundation	relief wells, drain materials, cut-off	Ahmed F. Chraibi
37	Morocco	Oued El Makhazine (culvert joint)	Embankment	Rockfill with central clay core	Seepage along conduit	turbid leakage, geotextile, culvert joints	Ahmed F. Chraibi
38	Morocco	Tuizgui Ramz	Gravity	Masonry	Seepage, internal erosion in foundation	leakage, collapse, monitoring system, erosion	Ahmed F. Chraibi
39	Norway	Muravatn	Embankment	Rockfill with central moraine till core	Pore water pressures in foundation	monitoring, fault zone, pressure, drainage gallery	Nilsen and Lien
40	Norway	Storvatn	Embankment	Rockfill with central inclined asphaltic concrete core	Settlements, deformation	asphaltic concrete core, membrane, stress, strain	Høeg, K

(Continued)

No.	Country	Dam name	Type	Material/shape	Issue	Keywords	Author
41	Norway	Svartevann	Embankment	Rockfill with central moraine till core	Pore water pressures, deformation	surveillance, monitoring, instrumentation, internal erosion, settlement, pore pressure, leakage, riprap	Goranka Grzanic
42	Norway	Trial dam	Embankment	Rockfill with central bitumen membrane	Impervious core	bitumen, membrane, pressure, deformations	Kjærnsli and Sande
43	Norway	Viddalsvatn	Embankment	Rockfill with central glacial moraine core	Instability of filter materials, internal erosion through dam	surveillance, monitoring, instrumentation, internal erosion, piping, leakage, sinkhole	Goranka Grzanic
44	Norway	Zelazny Most (Poland)	Tailings	Ring shaped	Pore water pressures, deformation	tailings dam, upstream method, geotechnical engineering	Jamiolkowski, M. et al.
45	Portugal	Automatic monitoring system	Concrete dams	-	Diligent monitoring data analysis	automated monitoring, dam operation, deformation measurement, analysis, safety of dams	Carlos Pina, Juan Mata
46	Romania	Gura Râului	Buttress dam	Concrete	Foundation rock deformation	drainage, surveillance, temperature, model	The Local Water Dam Administration of Olt
47	Romania	Paltinu	Arch	Double curvature concrete	Seepage through foundation, foundation rock deformation	drainage drills, seepage control, grout curtain	I. Asman
48	Romania	Pecineagu	Embankment	Concrete faced zoned rockfill	Cracks in concrete slabs, seepage	monitoring, geomembrane, leakage	A. Constantinescu
49	Romania	Poiana Uzului	Buttress	Round headed concrete	Seepage through foundation, foundation rock deformation	drainage, seepage, cracks, monitoring system, temperature,	Al. Constantinescu

(Continued)

	Country	Dam	Type	Material	Phenomenon	Topics	Author
50	South Africa	Belfort	Multiple buttressed arch & embankment	Concrete & homogeneous earthfill	Foundation rock deformation, concrete cracking	dam failure, foundation, crack pattern	Chris Oosthuizen
51	South Africa	Driekoppies	Embankment & gravity spillway	Zoned earthfill with central clay core & concrete	Pore water pressure, hydraulic fracture	monitoring, pressure cells, drilling, pneumatic drills	Chris Oosthuizen and Louis Hattingh
52	South Africa	Inyaka	Embankment & trough spillway	Zoned earthfill with central clay core & concrete	Hydraulic fracture	monitoring, total pressure cells, pore pressure, installation	C Oosthuizen & LC Hattingh
53	South Africa	Katse	Arch	Double curvature concrete	Foundation rock deformation	monitoring, instrumentation, pendulums, cracks, rockslide	Lithatane Matete & Chris Oosthuizen
54	South Africa	Kouga	Arch	Double curvature concrete	Ageing of concrete (AAR attack)	monitoring, instrumentation, concrete swelling, finite element model	Chris Oosthuizen and Louis Hattingh
55	South Africa	Ohrigstad	Embankment	Concrete faced rockfill	Seepage through dam	bio-location, leakage, flow path, divers	Chris Oosthuizen
56	Spain	La Aceña	Arch gravity	Concrete	Dam movements	monitoring, DGPS system, deformations	David Galán Martín
57	Spain	Caspe II	Embankment	Zoned earthfill with central clay core	Seepage, internal erosion in foundation	foundation treatment, incident detection, internal erosion, monitoring, piezometer, pore pressure, dam safety, seepage	René Gómez, Manuel G. de Membrillera
58	Spain	Cortes	Arch gravity	Concrete	Reservoir slope sliding	monitoring, landslide, quarry, excavation	Juan Carlos Elipe

(continued)

#	Country	Dam	Dam type	Material	Topic	Description	Authors
59	Spain	La Minilla + El Gergal	Gravity & arch gravity	Concrete	Improvement of monitoring systems	upgrade monitoring, amortization period, analysis of alternatives	F. Vázquez
60	Spain	La Loteta	Embankment	Zoned earthfill with central clay core & upstream clay blanket and cut-off wall	Seepage, internal erosion in foundation	cutoff wall, drainage, embankment dam, emergency plan, erosion, foundation treatment, impervious blanket, incident detection, internal erosion, karst, monitoring, piezometer, pore pressure, repair, dam safety, seepage, upstream blanket	René Gómez, Manuel G. de Membrillera
61	Spain	Siurana	Gravity	Concrete	Uplift	monitoring, safety, piezometer network, drains, drainage network, uplift pressure, rehabilitation, remedial work, surveillance	A. Vaquero and C. Barbero
62	Spain	Val	Gravity	RCC	Uplift, deformations	deformation measurement, foundation, monitoring, operation, dam safety, seepage, settlement, temperature, uplift	René Gómez, Manuel G. de Membrillera
63	Spain	55 dams in Ebro river basin	All dam types	-	Diligent monitoring data analysis	automated monitoring, safety of dams, statistical method	S. Hoppe, R. Gómez López de Munain and Manuel G. de Membrillera Ortuño
64	Sweden	Storfinnforsen	Embankment	Zoned fill with central glacial till core and timber sheet pile	Seepage, internal erosion in dam	pore pressure rise, timber sheet pile, leakage, geotechnical investigations, temperature, optical fiber, stabilizing berm, drainage system	Sam Johansson, Carl-Oscar Nilsson, Ake Nilsson
65	Sweden	20 high risk dams	Several dam types	-	Improvement of monitoring systems	behaviour, monitoring, training, potential failure mode analysis	Sam Johansson & Pontus Sjödahl

(continued)

66	Tunesia	Ziatine	Embankment	Zoned earthfill with central clay core	Seepage, internal erosion in foundation and contact	drainage, embankment dam, hydraulic gradient, internal erosion, leakage, piping, pore pressure, remedial work, seepage	N.H. Dhiab
67	USA	Dorris	Embankment	Homogeneous earthfill	Seepage along conduit	potential failure modes, seepage along conduit, effective visual monitoring	J. Stateler and B. Iarossi
68	USA	Ochoco	Embankment	Homogeneous earthfill	Seepage through dam	potential failure modes, piezometers, seepage through ancient landslide	J. Stateler
69	USA	Steinaker	Embankment	Zoned earthfill with central clay core	Upstream dam slope sliding	potential failure modes, upstream slope failure, responding to incident	J. Stateler
70	USA	Wanapum	Gravity & embankment	Concrete & zoned earthfill	Sliding of dam body	potential failure modes, sliding in body of concrete dam, design calculation error	K. Marshall
71	USA	Wolf Creek	Gravity & embankment	Concrete & earthfill	Seepage through foundation	karst foundation, seepage through foundation, water pressure data	B. Walker, J. Bomar, and V. Bateman

Hazard or potential failure mode	Malpasset (France)	Vajont (Italy)	Zeuzier (Switzerland)	Teton (USA)	Dnieprostri / Möhne / Eder (USSR / Germany)	Folsom (USA)	Cahora Bassa (Mozambique)	Zoeknog (South Africa)	Tous (Spain)	El Chocón (Argentina)	Durlassboden (Austria)	Gmuend (Austria)	Zillergründl (Austria)	Comoé (Burkina Faso)	Song Loulou (Cameroun)
Proper opertaion and maintenance							1		1						
Sabotage					1										
Diligent monitoring data analysis							1								
Improvement of monitoring systems							1		1				1		
Sedimentation															
Earthquake														1	
Reservoir slope sliding		1													
Downstream river erosion															
Dam slope stability										1					
Sealing membranes (cracks & behaviour)															
Ageing of concrete & cracking							1								1
Temperature load															
Sliding of dam body															
Uplift	1												1		
Settlements, deformations and movements of dam body							1								
Foundation rock deformation	1														
Hydraulic fracture								1							
Pore water pressures			1					1							
Erosion (foundation and dam body)				1						1	1			1	
Seepage				1						1	1			1	

(Continued)

DOI: 10.1201/9781003274841-8

	7	8	9	10	11	12	13	14	15	16	17	18	19	20
Proper opertaion and maintenance														
Sabotage														
Diligent monitoring data analysis														
Improvement of monitoring systems														
Sedimentation			1											
Earthquake														
Reservoir slope sliding														1
Downstream river erosion														
Dam slope stability											1			
Sealing membranes (cracks & behaviour)	1												1	
Ageing of concrete & cracking												1		
Temperature load														
Sliding of dam body														
Uplift				1	1		1							
Settlements, deformations and movements of dam body									1					
Foundation rock deformation												1		
Hydraulic fracture														
Pore water pressures														
Erosion (foundation and dam body)				1	1		1	1		1				
Seepage	1	1		1		1		1					1	
Dam	Kootenay Canal Forebay	WAC Bennett (trends analysis)	WAC Bennett (ROV)	Not reported	Not reported	Porce II (assessment study)	Porce II (thresholds)	Santa Rita	Tona	Urrá I	Mšeno	El-Karm	Etang	Grand'Maison
Country	Canada	Canada	Canada	Canada	Canada	Colombia	Colombia	Colombia	Colombia	Colombia	Czech Republic	Egypt	France	France
N°	7	8	9	10	11	12	13	14	15	16	17	18	19	20

(Continued)

N°	Country	Dam	Seepage	Erosion (foundation and dam body)	Pore water pressures	Hydraulic fracture	Foundation rock deformation	Settlements, deformations and movements of dam body	Uplift	Sliding of dam body	Temperature load	Ageing of concrete & cracking	Sealing membranes (cracks & behaviour)	Dam slope stability	Downstream river erosion	Reservoir slope sliding	Earthquake	Sedimentation	Improvement of monitoring systems	Diligent monitoring data analysis	Sabotage	Proper operation and maintenance
21	France	La Paliere	1											1								
22	France	Mirgenbach																				
23	Germany	Sylvenstein	1	1																		
24	Iran	Alborz			1	1		1														
25	Iran	Gotvand			1	1		1														
26	Iran	Karun 4										1										
27	Iran	Masjed-e-Soleiman			1			1														
28	Iran	Seymareh																				
29	Italy	Ambiesta									1											
30	Italy	Isola Serafini										1										
31	Italy	San Giacomo							1						1							
32	Italy	MISTRAL																				
33	Japan	Kyogoku Upper Reservoir	1																	1		
34	Japan	Okuniikappu															1					
35	Japan	Yashio															1					
36	Morocco	Oued El Makhazine (cutoff wall)			1																	

N°	Country	Dam	Proper operation and maintenance	Sabotage	Diligent monitoring data analysis	Improvement of monitoring systems	Sedimentation	Earthquake	Reservoir slope sliding	Downstream river erosion	Dam slope stability	Sealing membranes (cracks & behaviour)	Ageing of concrete & cracking	Temperature load	Sliding of dam body	Uplift	Settlements, deformations and movements of dam body	Foundation rock deformation	Hydraulic fracture	Pore water pressures	Erosion (foundation and dam body)	Seepage
37	Morocco	Oued El Makhazine (culvert joint)																				1
38	Morocco	Tuizgui Ramz																			1	1
39	Norway	Muravatn																		1		
40	Norway	Storvatn																				
41	Norway	Svartevann															1			1		
42	Norway	Trial dam										1					1					
43	Norway	Viddalsvatn																			1	
44	Norway	Zelazny Most (Poland)															1			1		
45	Portugal	gestBarragens																				
46	Romania	Gura Râului																1				
47	Romania	Paltinu																1				1
48	Romania	Pecineagu																				1
49	Romania	Poiana Uzului										1						1				1
50	South Africa	Belfort											1						1	1		
51	South Africa	Driekoppies																	1			
52	South Africa	Inyaka																				
53	Lesotho	Katse																1				

140

N°	Country	Dam	Seepage	Erosion (foundation and dam body)	Pore water pressures	Hydraulic fracture	Foundation rock deformation	Settlements, deformations and movements of dam body	Uplift	Sliding of dam body	Temperature load	Ageing of concrete & cracking	Sealing membranes (cracks & behaviour)	Dam slope stability	Downstream river erosion	Reservoir slope sliding	Earthquake	Sedimentation	Improvement of monitoring systems	Diligent monitoring data analysis	Sabotage	Proper opertaion and maintenance
54	South Africa	Kouga										1										
55	South Africa	Ohrigstad	1																			
56	Spain	La Aceña	1	1																		
57	Spain	Caspe II						1														
58	Spain	Cortes														1						
59	Spain	La Minilla + El Gergal																	1			
60	Spain	La Loteta	1	1																		
61	Spain	Siurana							1													
62	Spain	Val						1	1													
63	Spain	55 dams in Ebro basin (DAMDATA)																				
64	Sweden	Storfinnforsen	1	1																1		
65	Sweden	20 high risk dams																				
66	Tunesia	Zlatine	1	1															1			
67	USA	Dorris Dam	1																			
68	USA	Ochoco	1																			
69	USA	Steinaker																				
70	USA	Wanapum								1				1								
71	USA	Wolf Creek	1																			

6.4. BENCHMARK CASE HISTORIES

- Malpasset Dam (importance of monitoring engineering geological aspects)

- Vajont Dam (importance of monitoring reservoir slopes)

- Zeuzier Dam (The unbelievable effect of pore pressure relief)

- Teton Dam (value of diligent visual observations)

- Dnieprostroi, Möhne and Eder Dams (explosive loads twice during World War II)

- Folsom Dam (gate failure…tested regularly but not all the way)

- Cahora Bassa Dam (the value of diligent installations on the life of instruments)

- Zoeknog Dam (failure…predicted by pore pressure gauges, but ignored)

- Tous Dam (backup systems failure)

DOI: 10.1201/9781003274841-9

MALPASSET ARCH DAM FAILURE AT FIRST IMPOUNDMENT

B. Goguel, France

ABSTRACT Case history category: e, b, f. The failure of Malpasset Dam was a watershed case history that benefitted dam-engineering in more ways than one. It highlighted the utmost importance of dam safety surveillance as well as the significance of proper geological- and geotechnical investigations.

TECHNICAL DETAILS

Malpasset Dam was built between April 1952 and October 1954. The dam is located near Fréjus, an old roman city on the Côte d'Azur, along the Mediterranean Sea, at a place called Malpasset (a bad pass for passing people). The 66.5 m high arch dam with a crest length of 222 m, was constructed in a narrow rocky section on a small river. Using only 48 000 m³ of concrete a reservoir with a capacity of 50 million m³, was created for irrigation and flood damping. The owner was the French State (Var Département, without any dam specialist at the time). For more details on the dam see Figure 1 to Figure 3.

The owner relied on the Engineer (André Coyne) for the design and for construction supervision, with a loose link for evaluation after the end of construction. Due to budgetary constraints no detailed geotechnical investigations were carried out before construction. The geology can be described as Gneiss (old metamorphic horst), very heterogeneous and crisscrossed by joints of all scales and directions.

SURVEILLANCE DETAILS

Customary at the time, no drainage curtain was provided; no organized leakage and no piezometric pressures were monitored. Visual inspections were only performed by the water bailiff of the dam. Geodetic measurements (of 27 targets on the dam d/s face) were performed in few yearly (summer) measurements (campaigns), as shown in Figure 4. After dam completion, reservoir impoundment was delayed due to lack of expropriation upstream of the dam, and lack of the water distribution network downstream of the dam. Flash floods in November 1959 filled the reservoir rapidly.

Figure 1
Malpasset Dam at the end of construction, mid 1954 (left) and soon after failure in December 1959
(right)

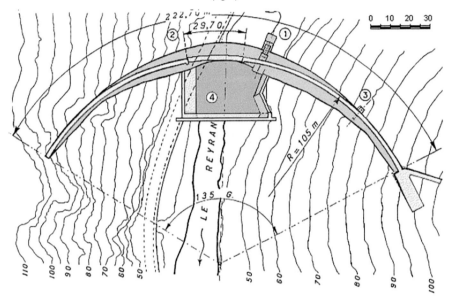

Figure 2
Plan view with contours in meters above sea level (1 - Bottom outlet; 2 -Uncontrolled spillway;
3 - Outlet works; 4 - Stilling basin)

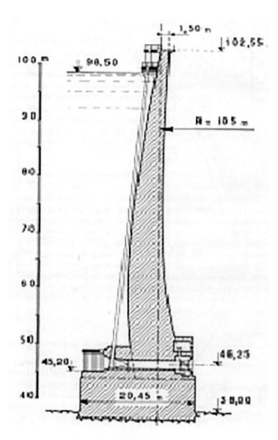

Figure 3
Highest cross section (the support of the 1.5 m diameter bottom outlet, fitted with
a downstream hollow valve and an upstream control gate, explains the
widened foundation at this cross section only)

Crest elevation	102.55	Normal operating level	98.50
Maximum flood level	102.00	Axis of outlet pipe	46.25
Spillwzy weir level	100.40	Lowest bedrock level	38.00

DESCRIPTION OF THE INCIDENT

On 2 December 1959, following intense rains and resulting floods, the reservoir level rose rapidly approaching the uncontrolled (non-gated, free sill) spillway level. The bottom outlet was opened after a meeting held on site that afternoon. During that meeting none of the 10 persons present noticed or reported any signs of distress on the dam or its abutments.

Results from the last geodetic measurement taken (see point D on Figure 4, July 1959, RWL 94.10 i.e. 8 m below the Max Flood Design Level) were not yet in the engineer's hands. (Afterwards, he could not say what would have been his comments on excessive movements in the lower sections of approximately 10 mm).

The dam failed the same evening around 21:11 after a sudden failure of the left bank. A large wedge of rock dislodged, below the left half of the arch dam, leaving an excavation in a dihedral form as shown in Figure 5. Only the bases of the right bank and central concrete blocks were still there, as shown in Figure 6. All other parts of the arch washed away; some blocks of concrete were swept away for more than 1.5 km.

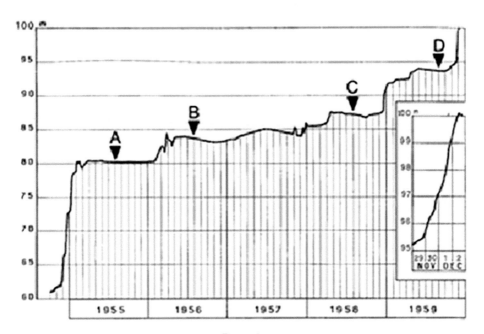

Figure 4
Reservoir level during years 1954–1959 (the box magnifies the four last days).
Triangles = dates of geodetic measurements A (late reference) to D.

Figure 5
The left bank after failure. "Dihedral" excavation with half of the upper thrust block fallen after the failure (the exploratory adit on the lower right was excavated after the failure to perform in situ jack tests).

The displacements of the remaining parts correspond to a rotation of the whole dam as a solid body, the axis of which is located at the very end of the arch at the right bank (and inclined perpendicularly to the fault plane forming the d/s face of the rock wedge which supported the dam and disappeared on the left bank). This resulted in an open crack at the upstream heel, 0.5 m open at the crown.

The upstream face of the wedge comprised of several shear planes parallel to the foliation of the gneiss, dipping towards the river. During the design and construction stages the downstream fault was not known. It was between 15 and 40 m below the lowest excavations (Figure 7), and it surfaced unnoticed in the overburden, more than 20 m downstream of the dam.

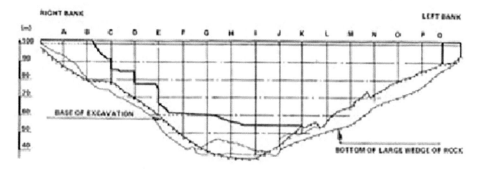

Figure 6
Downstream developed elevation after the failure

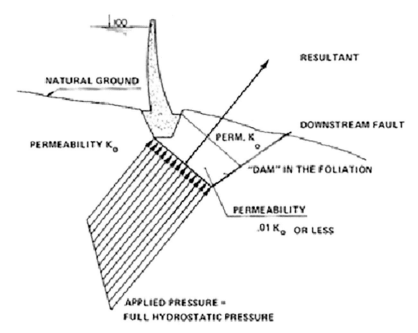

Figure 7
Full hydrostatic water pressure in the left flank foundation of the dam

After this catastrophic failure (resulting in 423 casualties and severe losses), several conventional back-analyses were undertaken. However, none could lead to an explanation for the failure:

- concrete stresses were moderate and well below what could fail the dam,

- buckling of the arch was not realistic,

- sliding of the abutment block was not the initial cause, rather a consequence of the overall rotation of the arch,

- sliding of the arch at contact with foundation was also discarded,

- sliding on the downstream fault was not possible, unless high pore pressures existed at depth in the rock (much in excess of what could be expected from a conventional flow net analyses).

Research studies therefore focussed on the seepage flows in the foundation. These studies contributed to a better understanding of "rock hydraulics". Seepage in rock masses, and the internal forces it generates, is governed by the main characteristics of the rock masses, which are discontinuous, non-homogeneous and anisotropic. Several of these new findings finally contributed to explaining the failure mechanism:

- rock permeability may be highly sensitive to stress, depending on the degree of jointing and the nature of joints infilling; geologic and petrographic studies established that the intensity and nature of the fissuration and jointing at Malpasset Dam occurred at all scales; the load applied by the dam to its foundation was therefore able to induce a drastic reduction of permeability under the structure,

- stress in discontinuous media can penetrate quite long-distance following discontinuities, particularly in the direction of the rock foliation, therefore forming a watertight barrier at depth under the dam toe,

- upstream crack developing in the arch dam foundation provided a preferential path for the development of full hydrostatic pressure, which could penetrate deeply when combined with the above-mentioned watertight barrier,

- deformability of the foundation was relatively high, with very low modulus of elasticity in the left bank (as demonstrated by finite element back analyses performed during the eighties).

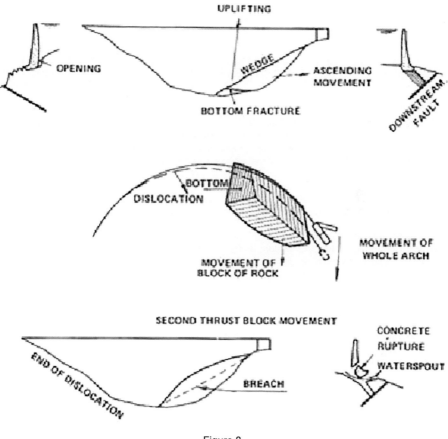

Figure 8
Schematic representation of the "explosive" failure phase

The watertight barrier, generated by the extremely sensitive rock of Malpasset Dam in the direction of the dam thrust, was located deep in the left bank where it coincides with the foliation planes. The development of high hydrostatic water pressures (uplift) consequently made sliding on the downstream fault possible. (At the right bank, the watertight barrier could not form in the same location because the foliation was not symmetrical with respect to the dam).

A failure scenario could therefore be reconstructed based on these observations. This has also led to the development of the Londe method for 3D stability analysis (which became a standard in arch dam design and safety studies).

LESSONS LEARNT

The concept of appropriate dam safety-monitoring using reliable and well-located instruments has become the standard. This was critically absent at Malpasset Dam. Piezometric measurements in the foundation together with strain meters in the arch, and monitoring of the foundation deformations, would probably have given ample warning at least several months prior to the failure.

The importance of rapid transmission of appropriate monitoring information to experienced engineers was emphasised. The geodetic measurement system at Malpasset Dam was monitoring arch deformations rather than foundation movements (due to the reference measurements for the computational models not been taken). The calculation delay at Malpasset Dam was also excessive; as well as the lack of a chain of suitable personnel to evaluate the information.

Instruments for the precise and frequent foundation deformation monitoring have since been developed. Inverted plumblines, for example, were not yet available during the construction of Malpasset Dam. They only appeared on the market during the very late fifties - early sixties and are nowadays common use everywhere for the monitoring of foundations.

In the absence of preconceived potential failure mechanisms, instrumentation may not always provide the required information. It is therefore important that monitoring (instruments and their reading frequency) be designed pro-actively to monitor potential failure mechanisms.

Deep drainage curtains became a standard feature in arch dams (especially the flanks). Malpasset Dam opened the eyes of dam designers on uplift acting not only below and inside the structure, but also inside the rock mass at depth and downstream of the dam.

Rock mechanics applied to dam engineering was developed intensively after the failure of Malpasset Dam. Systematic detailed geological and geotechnical investigations protocols for new projects have also been developed internationally.

Pro-active surveillance of dams during their life cycle and particularly during first impoundment became the standard. More generally, the value of an Independent Panel of Experts different from the Owner and Engineer has been emphasized.

SELECTED REFERENCES

Londe P., 1987, *The Malpasset dam failure*, Engineering Geology, Elsevier, Amsterdam, 24, 295–329.

Post G., Bonazzi D., 1987, *Latest thinking on the Malpasset accident*, Engineering Geology, 24, 339–353.

Carrère A., 2010, Les leçons de Malpasset, leur application aux projets de barrages d'aujourd'hui, Revue française de Géotechnique, Paris, 131–132, 37–51.

Duffaut P., *The traps behind the failure of Malpasset arch dam, France, in 1959*, Journal of Rock Mechanics and Geotechnical Engineering (2013), DOI.

THE VAJONT DISASTER

R. Gómez López de Munain & J. Fleitz, Spain

ABSTRACT Case history category: e, a, b. Construction of the Vajont Dam started in 1956 and was completed in 1960, at this time it was the highest double-curvature arch dam in the world – with a height of 261.6 m, a crest length of 190 m and a reservoir capacity of 168.7 million m^3 water. During the third filling, on 9 October 1963, Monte Toc, located on the left side, experienced a translational landslide which led to 270 million m^3 of rock falling into the reservoir, causing a 235 m high wave which passed over the dam by 100 m and arrived at the Piave valley, the wave sweeping away the downstream villages of Longarone, Pirago, Villanova, Rivalta and Fae, causing the death of 2 000 people.

DESCRIPTION OF THE DAM

Between 1957 and 1960, the Adriatic Electric Company (SADE) built the Vajont Dam, located 100 km north of Venice (Italy). At that time, it was the highest arch dam in the world and the second highest dam at 261.6 m. The crest length is 190.5 m, and the reservoir capacity 168.7 million m^3 (150 million m^3 of useful volume) with a normal maximum level of 722.50 m a.s.l. Approximately 360 000 m^3 of concrete was used in its construction. The dam was 3.4 m thick at its crest widening out to 27 m at its base. The designer and construction manager was Carlo Semenza, an experienced and internationally renowned civil engineer.

Figure 1
View of Vajont Dam before the event

SURVEILLANCE DETAILS

The first concerns about the stability of the valley slopes were raised during construction of the dam, with a number of reports being compiled during 1958 and 1959. These reports identified a possible prehistoric slide on the right bank. In view of the possibility of previous landslides and the synclinal form of the strata there was considerable discussion on the stability of the valley walls. In 1959, before the dam was finished, some field investigations were undertaken. A number of existing and potential landslides were identified, but the analyses of them suggested that the likelihood for large-scale movements was limited based on the following results of the field investigations:

- 3 test borings had failed to identify areas of weakness;

- It was assumed that any shear plane would have a 'chairlike' form and exert a 'braking effect';

- Seismic analyses had suggested that the banks consisted of very firm in-situ rock with a high Young's modulus.

As a result, no action was taken to stabilize the slopes, but a monitoring program was set up. In the following three years, the downward motion of the slide was monitored by means of surface markers. Some of the data provided by them are plotted in Figure 2. In addition, water pressures in perforated pipes, located in four boreholes (location shown in Figure 2), were monitored, starting in July 1961.

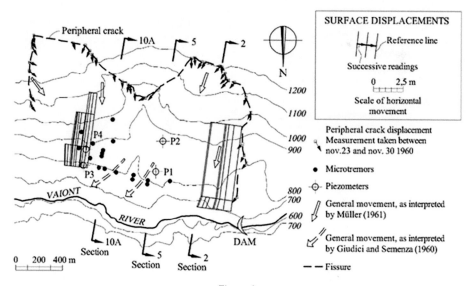

Figure 2
Map of the Vajont sliding area. Note the position (and comparative size!) of the arch dam on the lower right-hand corner of the figure (Alonso et al.: 2010)

ANALYSIS OF FIELD DATA DURING RESERVOIR IMPOUNDMENT

The dam was completed in 1960 and the impoundment of the reservoir started. In October 1960, with the reservoir partially impounded, a long, continuous peripheral crack, 1 m wide and 2.5 km in length opened and marked the contour of a huge mass, creeping towards the reservoir in the northern direction. The landslide was moving continually, around three to four cm/day.

Data on horizontal displacements, plotted as a function of position and time in several profiles following the south-north direction in Figure 2, suggest that the slide was essentially moving as a rigid body. The direction of the slide is also indicated in the figure by several arrows. Some of them (small arrows along the peripheral crack) indicate that the moving mass was actually detaching from the stable rock, implying no friction resistance along the eastern and western boundaries of the slide.

On November 4, 1960, when the reservoir level reached 180 m, a 700,000 m³ lump fell off the front of this larger landslide over a period of about 10 minutes. The speed of the slope displacement at this time was 3–4 cm/day (Oct 1960) with a maximum displacement of 100 cm.

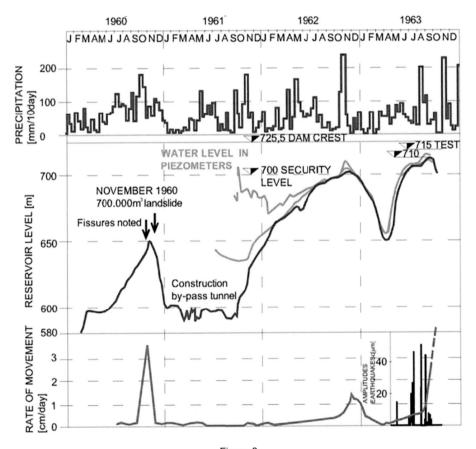

Figure 3
Summary of monitoring data recorded at Vajont Dam: precipitation, water levels in the reservoir and groundwater levels (measured with piezometers) and rate of movements (Belloni et al: 1987)

This event caught the attention of the site managers and they decided to lower the reservoir level down to 135 m and the movements reduced to close to 1 mm per day. They also tried to build a slope drainage tunnel but finally abandoned because it was impossible to progress and to sustain the tunnel due to the strongly fractured rock material.

At that time the designers realised that the large mass of the left bank was inherently unstable. However as there was no realistic way of arresting the slide artificially and stop the slide, or safely cause the mass to slide down all at once, it was decided to use varying levels to try and gain control of the sliding mass. Actually, based on the limited monitoring data available (Figure 3) a correlation of the reservoir level with the measured vertical and horizontal displacements of a number of topographic marks distributed on the slide surface was established (Figure 4).

The dam engineers believed that by carefully elevating the level, movement of the large landslide mass could be started and then controlled by altering the level of the reservoir. It was calculated that, should a sudden movement occur, providing it did not lead to a filling of the reservoir in under 10 minutes, it would not cause over-topping of the dam.

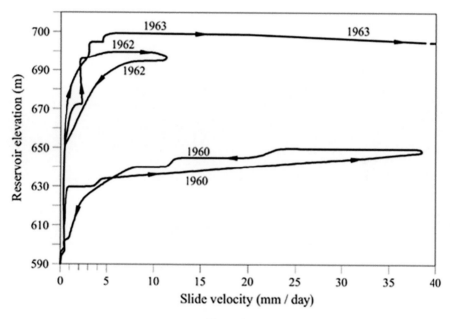

Figure 4
Relationship between water level in the reservoir and sliding velocity (Alonso et al: 2010)

As a result of such a "controlled" landslide the reservoir would be partially filled with the slope material. Therefore, a by-pass tunnel was constructed to connect the upper and the lower part of the reservoir to guarantee the future operation of the dam.

For most of 1961, the lake level was kept deliberately low to allow construction of the tunnel. In October of that year the site was ready once again and filling was restarted. The plan was to slowly fill the level of the lake while the movement of the landslide was monitored. The intention was that the landslide would slowly slip into the lake. If the movement rate became too high the reservoir level would be dropped to slow the movement down. Thus, the movement of the landslide would be controlled by varying the lake level.

From October 1961 to November 1962, the level of the lake was slowly increased. Late in 1962 the movement rate became too high, so the lake was partially emptied, whereupon the landslide effectively stopped. Filling was then restarted in April 1963. By early September, the water depth was 245 metres. The rate of movement of the landslide slowly increased, and in late September, the water level was slowly dropped in an attempt to reduce the rates of movement. The maximum level achieved was 710. The slope displacement speed was 1–3.5 cm/day. Despite the lowering of the reservoir level, the slope displacement continued to accelerate, reaching 20 cm/day the same day of the failure.

FAILURE

At 10:38 pm on October 9, 1963, the landslide collapsed. The entire mass of approximately 270 million m^3 of earth and rock slid 500 m north at up to 30 m sec-1 (110 km per hour). The huge velocity reached by the landslide magnified the tragic consequences. When the mass fell into the lake it had a level of 235 m and contained about 115 million m^3 of water. The landslide pushed a wave up the opposite bank 260 m above the original reservoir level and destroyed the lower part of the village of Casso and then over-topped the dam by up to 245m. An estimated 30 million m^3 of water fell on five villages decimating them. About 2 000 lives were lost. Despite this the dam was left structurally intact, only a metre of masonry was washed away at the top.

Figure 5
Photograph of Monte Toc taken by E. Semenza on 1 September 1959 with the construction crane for the dam to the right. The lower line corresponds to the existing paleo-landslide and the upper line to the 1963 landslide limit

Figure 6
Identical view taken by Daniele Rossi at the end of October 1963 just after the disaster

Figure 7
View of the municipality of Longarone before (left) and after (right) the event

INVESTIGATIONS OF THE FAILURE

The tragedy at Vajont Dam has spurred numerous investigations into the conditions triggering slope collapse, some of them are mentioned in the references.

It is now widely agreed that failure occurred along bands of clay within the limestone mass, at depths between 100 and 200 m below the surface (Hendron & Patton: 1985). These clay beds, 5 to 15 cm thick, represented planes of weakness, which, though sub horizontal at distances less than 400 m from the gorge, were farther away inclined at about 35° into the valley. The presence of the clay layer was not detected by the field investigations.

Interpretation of sliding risk was essentially made on the basis of reservoir elevation and surface displacements. There was also information on rainfall and on the levels of four piezometers. The "piezometers" were in fact open tubes which did not reach the level of the sliding surface and only provided average water pressures prevailing along their length. In addition, no direct information of the position of the failure surface and, in particular, on the type of material being sheared was available.

Identification of a landslide for the purposes of estimating its evolution and of defining any remedial measure requires information of a few key variables. Ideally, these key variables should also be used in the formulation of a mechanical model of the motion. Early knowledge of data concerning the basal failure surface (geometry, pore water pressure, type of material and drained strength parameters) would have been fundamental to build a conceptual and mechanical model for the slide. In the case of Vajont Dam, the observation that the slide velocity decreased when the reservoir level was reduced, irrespective of the absolute level of the water, provided a reservoir filling criterion which, finally led to the failure.

In some sense, an "observational method" was applied. The observational method, described by Peck (1969), requires the following ingredients: a) direct observation of a key variable describing the essential nature of the problem; b) a proper conceptual, analytical or computational model able to provide an estimation of the risk, in a general sense, for some threshold values of the key variables and c) a plan, defined in advance, to act in a specified manner when threshold values are exceeded. The key variables to be interpreted were the displacement rates of surface markers and the reservoir level.

The conceptual model was essentially given by the preceding observation, illustrated in Figure 4. The action in mind, in case of excessive displacement rate, was to reduce the water level in the reservoir. It was accepted, despite this strategy, that a full slide was a likely event and that the expected height of the generated wave was even estimated by model studies. However, the conceptual model was not based on any mechanical analysis of the slide. In addition, the reservoir level did not necessarily provide the actual pore pressures on the failure surface and the remedial plans were too simple and weakly connected with the complex mechanisms taking place within the slide.

LESSONS LEARNT

Vajont Dam marked a change in the emphasis dam designers placed on the stability of reservoir slopes. The Vajont Dam experience illustrates the following aspects:

- Pre-existing ancient landslides can be highly sensitive and can be reactivated when a reservoir is formed against them.

- The importance of a thorough understanding of the geology (lithology, structure, material properties) and hydrogeology of the reservoir rim.

- The importance of searching for recognising and evaluating precedent evidence for past instability in the reservoir basin.

- The significance of weak seams or layers in a slope, especially if inclined towards the reservoir.

- The need of sufficiently reliable models for geotechnical analysis with clear acknowledgement of their uncertainties and limitations.

- The effect of changing reservoir levels on slope stability.

- The significance of the joint effects of rapid reservoir level changes and the various influences of rainfall on slope stability.

- The value of reliable monitoring data, prompt data evaluation and appropriate responsive actions (emergency action plans).

- The difficulty in predicting time of failure, landslide velocity and subsequent wave size and consequences.

- Worst case scenarios must be taken into account.

Although the sliding risk of the reservoir slope had been correctly identified by the construction team, the assumptions regarding the magnitude, the detailed failure mechanism and therefore the estimation of the consequences were erroneous because they were made upon very limited field investigations and monitoring data. It should be borne in mind that this comment is made more than 60 years after the first investigations started at Vajont Dam. The purpose is to learn from the case, not to criticize the involved individuals who had to work with the techniques and rules of practice available at that time.

Even today, managing a very large landslide is a daunting task. It is relatively easy to extract field data (pore water pressures, absolute deformations, "insitu" tests) in the first tens of meters of soil and rock. Going beyond 200 m requires sophisticated, not easily available, and time-consuming efforts. A very large landslide requires a vast site investigation and is not a matter of only a few borings.

Since the 1960s, significant progress in terms of field investigations as well as modelling and computer calculations of slope stability has been made.

Monitoring technology applied to slope control has developed significantly, including inclinometry, radar interferometry, GPS control, automatic topography stations, etc. Real time measurements of interstitial pressures, displacements of the failure surface, superficial displacements, recording seismic accelerations on the slope, including distortions and stress transmitted to other structures is feasible today.

Emergency action plans are now mandatory in many countries and combined with adequate surveillance programs are crucial to minimize consequences in case of failure, especially human fatalities.

However, in the words of Carlo Semenza, the dam designer, "...things are probably bigger than us and there are no adequate practical measures... After so many fortunate works and so many structures... I am in front of a thing which due to its dimensions seems to escape from our hands..." (from a letter written in April 1961; two and a half years before the disaster)

SELECTED REFERENCES

ALONSO, E (2005), *Las Catástrofes y el Progreso de la Geotecnia*, Real Academia de Ingeniería de España.

ALONSO, E. ET AL. (2010), *Geomechanics of Failures. Advanced Topics*, DOI 10.1007/978–90-481–3538-7_2, © Springer Science+Business Media B.V. 2010

BELLONI, L.G. AND STEFANI, R. (1987) The Vajont slide: Instrumentation – Past experience and the modern approach. Engineering Geology 24, 445–474.

CARLONI, G.C. AND MAZZATI R. (1964). *Rilevamento geomorfologici della frana del Vajont*. Annali del Museo Geologico di Bologna.

GIUDICI F. AND SEMENZA E. (November 1960). *Estudio geológico sul serbatoio del Vajont*. Societá Adriatica di Electtricitá.

GUPTA, H.K. (1992), *Reservoir-Induced Earthquakes*, Elsevier.

HENDRON A.J. AND PATTON F.D. (1985). *The Vajont slide, a geotechnical analysis based on new geological observations of the failure surface*. Technical report GL-85-5, Department of the Army, US Army Corps of Engineers, Washington D.C. 2 Voll.

ICOLD BULLETIN 124 (2002). Reservoir landslides: investigation and management - Guidelines and case histories.

NONVEILLER, E. (1987), *The Vajont reservoir slope failure*, Engineering Geology, 24, pp.493–512.

PASUTO, A. AND SOLDATI, M. (1990). *Some cases of deep-seated gravitational deformations in the area of Cortina d'Ampezzo (Dolomites)*, The Proceedings of the European Short Course on Applied Geomorphology, 2, 91–104.

PECK, R. (1969). *Advantages and Limitations of the Observational Method in Applied Soil Mechanics*. Ninth Rankine Lecture, Géotechnique, June 1969, 19, pp. 171–187.

PETLEY DAVID N. (1996). On the initiation of large rockslides: perspectives from a new analysis of the Vajont movement record.

ROSSI D. AND SEMENZA E. (1965). Carte geologiche del versante settentrionale del Monte Toc e zone limitrofe, prima e dopo il fenomeno di scivolamento del 9 ottobre 1963, Istituto di Geologia dell Úniversita di Ferrara.

ROSSI D. AND SEMENZA E. (1968). *La bassa valle del Vajont e lo scivolamento del 9 ottobre 1963*. In Leonardi P et al. Manfrini, Rovereto, Vol II.

SEMENZA EDOARDO (2001). *La Storia del Vajont*. K-flash Editore.

SITAR, NICOLAS (1997). *Kinematics and Discontinuous Deformation Analysis of Landslide Movement.* II Panamerican. Symposium on Landslides, Rio de Janeiro, Nov. 13–14th, 1997. Tika, Th.E. and J.N. Hutchinson (1999), *Ring shear test on soil from the Vajont slide slip surface.* Géotechnique, 49, N°. 1, pp. 59–74.

TORÁN JOSÉ (1963). *Vajont. Notas previas a un memorandum técnico.* Revista de Obras Públicas número 2982.

URIEL ROMERO, S. AND R. MOLINA (1974), *Kinematic aspects of Vajont slide.* Proc. 3rd Int. Conf. ISMR, Denver, Vol. 1-B, pp. 865–870.

VALDÉS AND DÍAZ-CANEJA J.M. (1964), *Meditaciones sobre la catástrofe de Vajont*, Vol. N° 20 del Servicio Geológico del M.O.P., Madrid.

VARDOULAKIS, 1 (2002), Dynamic thermo-poro-mechanical analysis of catastrophic landslides, Géotechnique, 52, N°.3, pp. 157–171.

ABNORMAL BEHAVIOUR OF ZEUZIER ARCH DAM

Henri Pougatsch & Laurent Mouvet, Switzerland

ABSTRACT Case history category: e, a, b, c. In December 1978, unusual deformations were detected at the arch dam of Zeuzier. It was established that they were the result of a modification of the hydrogeological conditions due to the construction of the Rawil road tunnel exploration adit located 1.4 km from the dam. The latter was significantly disrupted by ground movements. After detailed investigations, the dam has been conditioned and made operational again in summer 1987. The Zeuzier dam mostly gave the opportunity to be cognizant of this kind of events.

TECHNICAL DETAILS

The 156 m Zeuzier arch dam, with a crest length of 256 m and elevation at 1777 m a.s.l. in Switzerland, was commissioned in October 1957. Its reservoir has a storage capacity of 51 million m^3 (see Figures 1 and 2). The seasonal storage is mainly intended for hydroelectric power, part of it being also used in winter by a nearby village for water supply. The dam is founded on a rocky local narrowing of a valley shaped by Jurassic limestones. The stiff limestone layers are subdivided by many cracks. The high permeability of the dam foundation makes the groundwater under the dam almost nonexistent. Below the limestones, a confined groundwater aquifer does exist in a Dogger-formation protected by an aquiclude. However, the dam site was considered as suitable for a high arch dam.

The dam behaved satisfyingly until fall 1978.

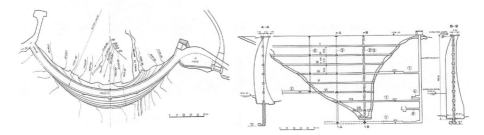

Figure 1
Plan, elevation and cross section

Figure 2
Downstream view, upstream view with empty reservoir

SURVEILLANCE DETAILS

The dam was very well equipped with surveillance instrumentation: 3 direct pendulums, survey of measurement targets on the downstream face, thermometers in the dam body and seepage and drainage flow measurement weirs. All measurements were recorded on a monthly basis. The dam behaved as predicted, and the statistical analysis carried on clearly demonstrated the reversible movement of the body of the dam due to variation of hydrostatic and thermal loads.

DESCRIPTION OF THE INCIDENT

At the beginning of December 1978, the monthly measurement of pendulums indicated that the deformations obviously differed from the former elastic behaviour of the structure. Easily and rapidly detected, the dam crown started to move toward the upstream direction although the reservoir was nearly full. In January 1979, movement reached more than 5 mm in both radial and tangential directions. End of March 1979, movement was more than 20 mm toward upstream and 15 mm toward the abutment.

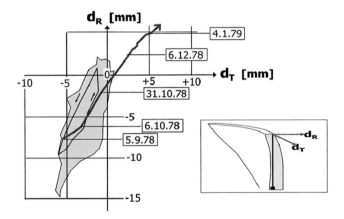

Figure 3
Radial movement at the crest, central pendulum, before and during the event [mm]

Green area: envelope for the period 1958–1976

Blue line: Oct. 77 to Sep. 78

Red line: since October 1978

Later, the analysis permitted to identify the beginning of the abnormal behaviour in the last days of September 1978. The reservoir was rapidly drowned down to its minimum water level and the frequency of monitoring measurements was increased.

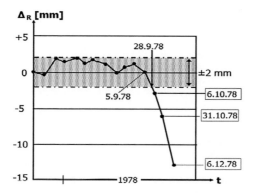

Figure 4
Radial movement at dam crest central section: variation between measurements and deterministic model (in red, considered acceptable variation: +/- 1 mm)

FOUNDATION MOVEMENT AND DAM DEFORMATIONS

Due to unfavourable weather conditions (heavy snow), the geodesic survey could only be carried out in spring 1979. The results indicated a settlement (almost 10 cm) and a movement towards the upstream (almost 9 cm) at the level of the crest as well as a narrowing of the abutment. Actually, the valley itself was narrowing (almost 7 cm). Further to these movements, opening of vertical joints in the top of the upstream facing and the development of cracks in the downstream facing has been noticed.

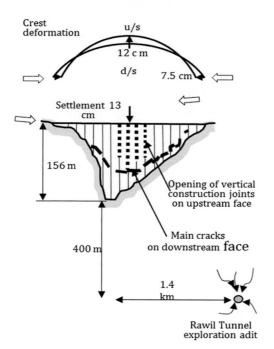

Figure 5
Maximum dam deformations and crack pattern observed on both faces. Width of main cracks

(December 1980): 1 to 15 mm

INVESTIGATION OF CAUSES - ROCK DEFORMATIONS

To determine the reasons behind the extraordinary dam deformations, the installed measuring system was significantly improved. Numerous instruments were installed (i.e., inclinometers, piezometers, extensometers) particularly in the foundations. The geodetic network was considerably extended, reinforced and adapted specially to make the measurements possible during the winter season.

The analysis of diverse hypotheses (mainly tectonic and hydrogeologic) pointed out that these unusual deformations followed a general bowl-shaped settlement of the ground extending on 2 to 3 km. The hypothesis of the influence of the drilling of a tunnel exploration adit for the construction of the Rawil road tunnel was immediately considered as the tunnel was drilled only 1.4 km away from the Zeuzier dam. This hypothesis was supported by a striking connection between the curve of the exceeding deformations of the dam and that of the cumulative water flows drained by the above-mentioned gallery. The establishment of a mathematical model representing the behaviour of a cracked elastic saturated rock massif scientifically confirmed that the drilling of the adit was at the origin of the settlement of the ground, (Lombardi. 1992, 2004). Consequently, the tunnelling works were interrupted in March 1979.

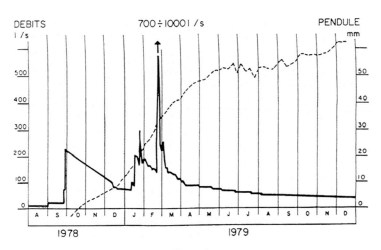

Figure 6
Flow in the Rawil tunnel exploration adit (full line) and deformations shown by the pendulum (dotted line)

REFURBISHMENT WORKS

The rehabilitation of the dam could begin only when some major conditions were fulfilled: (1) the movement of the ground and of the dam have practically stopped. (2) the capacity of the dam to be refurbished is ensured (i.e., the state of the foundations is checked, the ability to restore monolithism is verified and the stress-strain conditions are controlled) and (3) a decision is made on the future works of the tunnel exploration adit.

Once these conditions were met, the refurbishment works were applied in three different zones:

- The grout curtain: the investigations showed that the rock was of good quality and the grout curtain was still efficient.

- The rock-concrete contact zones: based on drill holes, it was possible to detect the zones where the rock-concrete contact had to be recovered.

- The dam body works in this area concerned the treatment of the cracking and the vertical joints. Specific tests were undertaken in the lower part of the dam and allowed to choose adequate epoxy resin and injection procedure for the crack and joint grouting.

The works were successfully executed between 1982 and 1984.

REFILLING OPERATION

Due to the damages of the dam and its foundations, a schedule for the progressive and controlled refilling was prepared. This schedule consisted on 6 annual steps. It included stops to perform intermediate measurements allowing to compare with previous steps. For each step, directives such as the kind of measurements to perform, visual inspections, etc. were previously defined in a document.

The filling operation was successfully achieved, and the normal water level was reached again by the summer of 1987.

LESSONS LEARNT

The organization of dam surveillance

The events of the Zeuzier dam highlighted the importance and appropriateness of the multi-level dam surveillance prescribed in Switzerland. It includes the owner's operation staff (i.e. measurement, visual control and operational tests), a civil engineer (analysis of the measured data, control of the structure's state), experts in civil engineering, geology and geodesy, as well as the high surveillance Authority.

It should be noted that the visual inspections were considered as crucial parts of the surveillance (SCD, 1997). It is also of great importance to create an archive file for the whole dam and to regularly update it (SCD, 2005). A monograph can also be used to summarize the essential events in the lifetime of the dam (SCD, 2002).

Monitoring system and analysis of the results

The case of Zeuzier clearly underlined the importance of the continuous supervision, the implementation of a large monitoring system as well as the fast analysis of the results of the measurements. It also highlighted the required means to be implemented (particularly the geodetic tools) in order to detect, in time, any abnormal behaviour and explain the causes. The monitoring system has to be designed to take into account measurements of loads and its consequences on the dam, foundations and surrounding area.

In the concrete body of the dam, the pendulum is a simple and effective means. However, several other instruments can be placed at the foundations to detect structural deformation (e.g., sliding micrometers), seepage flows as well as drainage and infiltration. Geodesy was also very useful to explain the deformations observed at Zeuzier. Nevertheless, this kind of specialized measurements requires certain precautions particularly when performing the measurements during winter. Within the development of the geodetic measurement, the GPS system offers an elegant method to integrate stable geological points into the control network outside of the zone of deformation.

Analysis of the results

While the analysis of raw measurements of deformations allows to have good insight on the nature of the behaviour of the dam, the comparison between foreseen deformation and observed deformation offers the possibility to refine this analysis. The difference has to be limited within a given margin of tolerance if the behaviour is regular (SCD, 2003).

Incidence of underground works

Underground works such as galleries or tunnels can influence the hydrogeological conditions due to drainage in an important zone. Consequently, if such works are planned at a maximal distance of 3 to 4 km from a dam, precautionary measures must be taken (e.g., extension of the monitoring system, increase of the frequency of the measurements). For example, such measures were taken during the drilling of the Alpine transversals of Lötschberg and Gotthard. Deformations were noticed at the Nalps dam (close to the Gotthard Massif), which confirms the rightness of these measures.

SELECTED REFERENCES

BIEDERMANN, R. ET AL. 1982. Comportement anormal du barrage-voûte de Zeuzier. Abnormal behaviour of Zeuzier arch dam. *wasser energie luft – eau énergie air, Special issue to the 14th ICOLD Congress, Rio de Janeiro, Brazil,* 1974(3): 65–112. Baden, Switzerland.

LOMBARDI G. 1992, "The F.E.S. rock mass model - Part 1", Dam Engineering, Vol. III, Issue 1, February 1992, pp. 49–72, "The F.E.S. rock mass model - Part 2", Dam Engineering, Vol. III, Issue 3, August 1992, pp. 201–221.

LOMBARDI, G. 1996. Tassement de massif rocheux au-dessus de tunnels. Symposium Géologie AlpTransit, Zurich février 1996.

LOMBARDI, G. 2004. Ground Water induced settlements in rock masses and conséquence for dams. IALAD Integrity Assessment of Large Concrete Dams. Conférence in Zurich

POUGATSCH, H. 1990. Le barrage de Zeuzier. Rétrospective d'un événement particulier. *wasser energie luft – eau énergie air,* (82)9: 195–208. Baden. Switzerland (in French).

POUGATSCH, H. & MÜLLER, R.W. 2002. Alp Transit und die Talsperren. Sicherheit ist oberstes Gebot. *wasser energie luft – eau énergie air,* (94)9/10: 273–276. Baden, Switzerland (in German).

SCD. 1997. Surveillance de l'état des barrages et check lists pour les contrôles visuels. *Comité suisse des barrages – Swiss Committee on Dams,* Switzerland (in French).

SCD. 2002. Comité suisse des barrages – Swiss Committee on Dams : Monographie de barrage. Recommandation pour la rédaction (in German, French, Italian).

SCD. 2003. Comité Suisse des Barrages – Swiss Committee on Dams : Methods of analysis for the prediction and the verification of dam behaviour. *wasser energie luft – eau énergie air, Special issue to the 21th ICOLD Congress, Montréal, Canada* (95)3/4: 73–110. Baden, Switzerland.

SCD. 2005. Comité suisse des barrages – Swiss Committee on Dams : Dossier de l'ouvrage d'accumulation. Recommandations (in German, French, Italian).

SCD. 2006. Comité suisse des barrages – Swiss Committee on Dams : Dam Monitoring Instrumentation / Dispositif d'auscultation des barrages (Part 1 and 2). *wasser energie luft – eau énergie air,* (98)2: 143–180. Baden, Switzerland.

SCD. 2013. Comité suisse des barrages – Swiss Committee on Dams : Geodäsie für die Ueberwachung Von Stauanlagen – Empfehlungen für der Einsatz der geodätischen Deformationsmessung bei Stauanalgen.

SCD. 2015. Comité suisse des barrages – Swiss Committee on Dams : Role and Duty of the Dam Wardens. Level 1 surveillance of water retaining facilities. *Special issue to the 25th ICOLD Congress, Stavanger, Norway.*

POUGATSCH H. ET AL. 2011. Improvement of safety of Swiss dams on the basis of experience. Proceedings International Symposium on Dams and Reservoirs under changing Challenges, Symposium at ICOLD Annual Meeting Lucerne, pp. 145–152

TETON DAM FAILURE

J. Fleitz, Spain

ABSTRACT Case history category: e. This article is mainly an adaptation of a paper published at the annual ASDSO conference in 2009 on the experience made with Fontenelle Dam, Ririe Dam and Teton Dam in the USA and its influence on technical, social and organizational innovation. The Teton Dam failed in June 1976 and originated subsequent changes of the dam engineering community as well as the public's perception of dam safety. Many countries adopted or reviewed guidelines and regulations in the field of design, construction control, surveillance and safety due to the experience of the Teton dam breach. KEY WORDS: Dam failure, seep, soil arching, hydraulic gradient, emergency action plan.

TECHNICAL DETAILS

The Teton Dam was designed to provide irrigation, flood protection, and power generation in the lower Teton region of southern Idaho in the USA. The dam was designed as a zoned earthfill dam with a maximum height of 123.4 m above the lowest point of excavation. The crest length was 945 m and there were 5 embankment zones:

- Zone 1 - Central Core

- Zone 2 - Upstream and Downstream material adjacent to Zone 1 and in a blanket under zone 3 in the river valley and abutments

- Zone 3 - Random fill downstream of zone 2

- Zone 4 - Upstream cofferdam, later incorporated into upstream toe of dam

- Zone 5 - Protective exterior upstream and downstream rockfill

Figure 1 shows a plan view of the embankment as it was constructed.

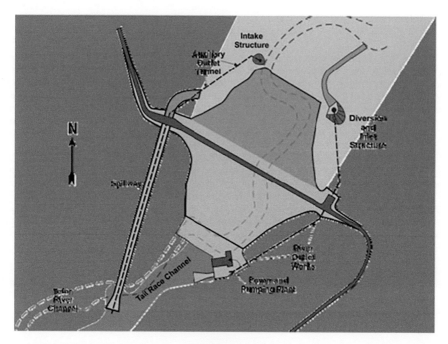

Figure 1
Plan view of Teton Dam

During the feasibility phase of the project, a pilot-grouting program was completed in the area of the key trench on the upper portion of the left abutment. The results showed that above El. 1554 m, the upper 20 m of rock was so permeable that blanket grouting was not practical from a cost standpoint. To compensate for the high grout losses in the pilot grouting program, a key trench was designed above El. 1554 m to connect the embankment core to the rock foundation.

To minimize costs, the trench was narrow - only 0.8 m wide at its base - and had sidewalls or side slopes of 0.5H:1V for most of its length. On the left and right abutments, the sidewalls are near vertical in some locations. The deep key trench was a first for Reclamation and won a design award for its cost-effective approach, which was later rescinded.

Construction of the dam began in February 1972 and was substantially completed in November 1975.

ENVIRONMENT AND FUNDING

Before construction of the dam could begin, a group of environmental organizations filed a complaint in Idaho District Court in 1971 to prevent construction of the dam. The lawsuit was dismissed from Federal District Court. The legal actions continued through 1974, with the plaintiffs alleging violations of numerous laws, including the National Environmental Policy Act of 1969 (NEPA). On December 23, 1974, the Ninth Circuit Court filed an opinion affirming the District Court's dismissal of the case, effectively ending the lawsuit. One of the main claims made by the plaintiffs was that the costs and benefits were misrepresented by the Government. The cost-benefit ratio for the Teton Dam project was 1.0:1.75, which is not a high ratio. Design engineers, construction forces, and other Reclamation employees who were associated with the Teton Dam project recall being constantly reminded about the extremely tight budget for the project. The likely source of the cost pressure was the cost-benefit ratio and the somewhat difficult process obtaining authorization in the face of litigation.

GEOLOGY

The Teton Dam site is adjacent to the eastern Snake River Plain, a volcanic filled depression that was formed by downwarping and downfaulting in late Cenozoic time. Older volcanic rocks are not exposed along the edges of the plain. The Teton River incised into a portion of the volcanic upland near the eastern end of the plain creating a steep-walled canyon at the dam site. The site is in an area of generally low seismicity. The foundation cut-off trench was excavated into bedrock along the entire length of the dam. The regional groundwater table is far below the river, though perched groundwater can be found above the channel.

The canyon walls are composed of a rhyolite welded ash flow tuff. The tuff was exposed in some areas, but talus slopewash and alluvial deposits predominantly cover it. The welded tuff is between 15 m to 180 m thick near the dam site and has prominent and abundant jointing intersecting and high and low angles. Most of the joints are near-vertical. The major joint set, strikes N25W to N30W, and is well developed on both abutments and in both outlet tunnels. A second joint set, striking N60W to N70W, is well developed in the lower upstream part of the right abutment, the river outlet works tunnel, and the downstream portion of the auxiliary outlet works tunnel. A minor set of northeast-trending, high-angle joints is also present in the welded tuff.

Continuous high-angle joints in the right abutment have been traced for lengths of as much as 60 m, but most are between 6 m and 30 m long. The aperture of most high-angle joints is less than 4 cm, but many joints are as much as several decimetres or even bigger. Examples of the jointing can be seen in Figure 2.

Figure 2
Right abutment key trench showing the jointed nature of the foundation

Many joints are open; others are partially or wholly filled with clay, silt, silty ash, soil, or rubble, especially near the natural ground surface. The permeability of the welded tuff is due entirely to the presence of open joints. The joints are most abundant and open, and rock-mass permeability is much higher above El. 1554 m. Many of the joints were infilled with erodible material that would soften on contact with reservoir water.

Underlying the welded tuff are materials of lacustrine, alluvial, and pyroclastic origin. Information about these materials has come mainly from drill holes, commonly with poor or no core recovery, and to some extent from deep grout holes, and thus is rather fragmentary. Although there is little information about the various units underlying the tuff, sand and gravel and variably cemented sandstone and conglomerate are commonly present. Thick claystone and siltstone are present under at least part of the left abutment and channel section. Thin ash-fall tuff and other pyroclastic materials were found below the welded tuff in some core holes. The contact between the sedimentary materials and the welded tuff is an irregular erosion surface with a local relief of at least 135 m and some slopes steeper than 30 degrees. The permeability of most of the sedimentary materials is less than that of the intensely ractured welded tuff, but is highly variable. The sedimentary materials are at least 120 m thick; the depth to the materials underlying these sediments is unknown.

Basalt is present in the bottom of the Teton River Canyon and is a remnant of a lava flow that filled the canyon to about El. 1526 m (see geologic section, Figure 3). In the dam foundation, the basalt is restricted to the left side of the river channel section, where it has a maximum thickness of about 38 m. It is separated from the underlying welded tuff by a deposit of alluvial material consisting of silt, sand, and gravel from 1.2 m to 6.7 m thick. The basalt is dense to moderately vesicular and contains closely spaced, randomly oriented joints and other fractures. In spite of its fractured nature, it is an adequate foundation rock for the dam. Water pressure tests showed the basalt to be tight and the thin alluvial fill between the basalt and the welded tuff to be permeable.

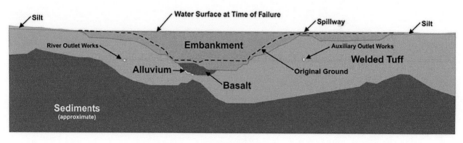

Figure 3
Profile of the Teton embankment and foundation looking downstream

FOUNDATION PREPARATION

The cut-off trench was excavated to competent rock. Foundation grouting was completed through a 0,9 m wide concrete grout cap that ran the length of the centerline of the cut-off trench. Overhangs were removed where possible, although excavations in the left abutment remnant found areas where overhangs were not removed, and abutment rock was not shaped. On the right abutment, shaping was limited because the location of the spillway prevented dramatic changes to the profile. No 'dental concrete' was used, and no slush grouting was used on the foundation surface above

El. 1585 m. Below El. 1585 m, 'dental concrete' was limited to using structural concrete leftover from placements elsewhere on the site, as there was no bid item for either dental concrete or foundation treatment. The rock surface treatment may have hastened the failure, because the criterion for treatment was not based on condition of rock, but on when excess concrete was available. 'Slush grout' was similarly taken from leftover foundation grouting materials.

Special compaction was done at the contact between the embankment and the foundation rock. During excavation of the left abutment remnant, it was discovered that special compaction was impossible to perform next to some of the open jointing. Foundation grouting was difficult given the extraordinary permeability of the Welded Tuff. Occasionally, structural concrete was poured in the large fissures and voids in the foundation where grout was considered inadequate. There was no consistent method to treat the rock foundations within the key trenches and there was no strong direction from designers as to how to treat the foundation. It is apparent that none of the designers knew how to adequately treat the foundation and that the organization did not understand the importance of doing so. Treating the foundation would also have meant ignoring the cost and schedule pressure from the construction office.

DESIGN CONSIDERATIONS

The design for Teton Dam followed standard practices with an impervious core and progressively coarser materials used as the zoning progressed downstream. In the maximum section, a relatively shallow-sloped cut-off leads to a narrow 9 m wide contact with a competent foundation. The intent of the design was to rely on the grout curtain and the cut-off to create an impermeable barrier to protect the main core of the dam. As shown in Figure 4, there are no additional defensive measures provided in either the main embankment or the abutment sections.

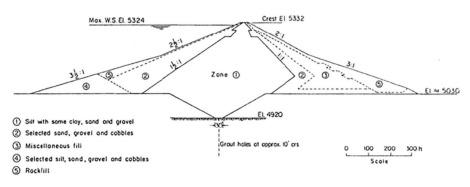

Figure 4
Teton Dam maximum section

On each abutment, the side wall of the cut-off trench steepens to 0.5H:1V. However, owing to the steepness of the abutment looking parallel to the axis, the actual sidewalls approach vertical on both abutments, which can easily be seen in Figure 5.

Figure 5
A view looking towards the left abutment from the right abutment

TETON DAM PERFORMANCE

On October 3, 1975, reservoir filing began; the reservoir was then at elevation 1542 m. From January 1, 1976 to April 5, 1976 the reservoir rose from elevation 1569 m to elevation 1577 m, or about 9 cm per day. From April 5 to June 1 the reservoir rose 36 m or about 58 cm per day. During the first 5 days of June, the reservoir level was increasing at an average rate of 70 cm per day.

The dam behaviour seemed to be normal, but when the reservoir level approached the spillway, on June 3, an inspection team downstream of the dam found clear water seeping from the ground at two locations: one at 400 m and another at 450 m downstream. By next day another seep had formed 40 to 60 m downstream. All three seeps formed downstream of the north side of the dam, but none were considered serious hazards.

Beginning at 7:00 a.m. on June 5, witnesses discovered the first seep on the dam itself and, shortly after, found a second seep. Turbid water flowed from both leaks on the north side of the dam. At 10:15 a.m., a wet spot formed and water began to leak out and erode the dam. Engineers recalled hearing a loud noise and rushing water at 10:30 a.m. Project Construction Engineer Robert Robison stated, "This leak developed almost instantaneous at about 10:30 a.m. and let loose with a loud roar". Within 10 to 15 minutes, two dozers began attempting to plug the hole and the County Sheriff was alerted to start downstream evacuations. Behind the dam, a whirlpool formed at 11:00 a.m. and grew rapidly despite efforts to dump rock in and seal the hole. At 11:30 a.m., drivers abandoned the two dozers on the front of the dam as the hole expanded and swallowed both dozers. One driver recalled running to get in another dozer to help with whirlpool efforts atop the dam. Fifteen minutes later these drivers were pulled back. At 11:57 a.m., the entire north embankment breached, and the nearly full reservoir came crashing through the earthen wall. The final report to the DOI presents a detailed hourly account of the dam's failure.

Although the time when the dam failed avoided the problem that lack of attendance would have caused, there was no one at the dam site from after midnight (12:30 a.m.) to 7:00 a.m. the day of the failure.

The dam failed with the reservoir at elevation 1616 m, 9 m below the dam crest. At the time of failure, the reservoir contained about 310 million m³ of water. A total of 296 million m³ of water drained from the reservoir within about 6 hours.

At the time the failure occurred, the power generating station was not yet complete and the auxiliary outlet works was under construction, limiting the release capacity of the facility during first filling. As the reservoir began to fill for the first time, a large amount of water pressure built up quickly in the lower portion of the cut-off trench on the right abutment probably between El. 1539 m and El. 1585 m. Untreated joints in the welded tuff in the upstream wall of the cut-off trench allowed full reservoir head to build up on the upstream face of the cut-off and an extraordinarily high gradient to be induced across the trench. Because of the narrowness of the cut-off trench, there were potentially low-stress areas due to soil arching. Soil arching in the right abutment key trench was demonstrated using finite element analyses. Following the investigations of the Independent Panel and the Interior Review Group, Jaworski, Seed and Duncan demonstrated by laboratory testing that soil on the upstream face of the key trenches that was adjacent to open joints would soften and collapse after wetting, allowing reservoir water closer to the grout cap and further increased the already enormous hydraulic gradient across the floor of the key trench.

Either in the lower portion of the cut-off or across the top of the grout cap, seepage from the high gradients began to erode material from the downstream face of the cut-off trench and carry it through the open jointed welded tuff to unprotected exits in the valley wall and through the pervious downstream embankment zones. Eventually, enough material washed out allowing erosion to progress upstream and creating an open pipe that connected to the upstream open joints and eventually connected to the reservoir. The pipe enlarged, progressed vertically and laterally, and breached the crest of the embankment causing a catastrophic failure and release of the entire pool. The uniqueness of the failure was the very rapid progression from discovery of a rather minor seep at the abutment to the total collapse of the dam. The speed of the progression to failure highlights the gross inadequacy of design and treatment to control seepage through the foundation

Figure 6
Photo sequence showing Teton Dam breach in roughly 90 minutes on June 5th, 1976

This embankment failure is the largest in the USA, based on the dam's structural height. 800 km^2 extending 130 km downstream were fully or partially inundated. 25 000 people were displaced, 11 people were killed, and the flood caused approximately $400 million in direct and indirect damage.

LESSONS LEARNT

Many interviewees from both the Corps and Reclamation expressed surprise that a dam could exhibit signs of distress at 7:00am and fully breach by 11:30am. Before Teton Dam, this seemed implausible to even experienced dam designers. The rapid nature of the failure and the vast devastation downstream changed both the dam engineering community and the public's perception of their own safety.

For these reasons, the Teton Dam failure has been considered as a benchmark case history, although even with the most appropriate surveillance and monitoring activities the dam break could not have been prevented due to its fast evolution during the first filling. The experience of Teton Dam lead to a complete review of the state of art and originated guidelines and regulations that among other aspects of design and construction control developed comprehensive safety and surveillance procedures. Many countries adopted or reviewed guidelines and regulations in the field of design, construction control, surveillance and safety due to the experience of the Teton Dam breach.

The key overarching conclusion from the failure is that the design of Teton Dam was not uniquely tailored to the site. The intense fracturing and open joints were not compatible with low-to-no plasticity silt. The left abutment grouting program should have been an indication of how susceptible the embankment might be to erosive forces. Instead of incorporating this into the design philosophy, the designers attempted to change the parameters of the problem by using the key trench in place of blanket grout.

Engineers that began their careers following the failure of Teton Dam started with an entirely new attitude that embraced state-of-the-art ideas, external review of decisions, and significantly more interaction between design and construction forces. Following the failure, Reclamation focused significant energy on education, and it thrived on in-house and external educational opportunities

The final lessons to be learned from the failure of Teton Dam are:

- Use more than one line of defense against seepage.

- Flaws can occur in man-made structures, and defense measures should be designed assuming that flaws do occur.

- External review of designs and decisions is a key step to evaluate the safety of a structure

- Critiquing problems and discussing controversial conditions is an important step to understand problems and planning the resolution to them.

- A central presence to facilitate communication between geologists, designers and construction forces is important.

- Foundation approval documented by designers and geologists for each square meter of material placed is an important consideration. Digital records of the foundation inspections should be required in the specifications.

- Incidents and failures should be openly discussed and presented as learning tools for all dam engineers.

- The decision structure of the organization must be continually observed and evaluated to see that effective decisions are being made.

- Communication with the downstream population is an important step to mitigate potential disasters.

Specific aspects regarding dam safety and surveillance that have been developed due to the experience of the Teton Dam failure are:

- The need for failure mode oriented and dam specific design of monitoring systems, especially the importance of an adequate seepage control in embankment dams

- Adequate and gradual first filling programs including specific inspection and monitoring activities to control and follow-up

- Fully operative discharge facilities to reduce the reservoir level if necessary

- Implementation of emergency action plans

SELECTED REFERENCES

SNORTELAND, NJ, SHAFFNER, P & DAVE PAUL, D (2009): Fontenelle Dam, Ririe Dam, and Teton Dam: An Examination of the Influence of Organizational Culture. Presented at Dam Safety 2009, held Sept. 27-Oct.1, 2009 in Hollywood, Florida.

IDAHO Bureau of Homeland Security. 1976 Teton Dam Collapse http://bhs.idaho.gov/Pages/History/DamCollapse.aspx.

GRAHAM, WJ (2015). The Teton Dam Failure - An Effective Warning and Evacuation. Presentation at the Association of State Dam Safety Officials 25th Anniversary Conference, Indian Wells, California, September 2015. http://damfailures.org/wp-ontent/uploads/2015/07/075_The-Teton-Dam-Failure.pdf.

ROGERS, JD & HASSELMANN, KF. Retrospective on the Failure of Teton Dam. http://web.mst.edu/~rogersda/teton_dam/Retrospective%20on%20Teton%20Dam%20Failure.pdf

INTENTIONAL DAM BOMBING ACTIONS DURING WORLD WAR II: MÖHNE, EDER AND DNIEPROSTROI DAMS

Pierre Choquet, Canada

ABSTRACT Case history category: e. This article collects publicly available information about intentional dam demolition actions on dams by air bombing raids or bombing actions during World War II.

MÖHNE AND EDER DAMS

The Royal Air Force of Great Britain conducted air bombing raids on a number of dams in the Ruhr region of Germany in the night of 16–17 May 1943.

Three dams were mainly targeted:

- Möhne Dam (masonry gravity completed in 1913, 40.3 m high, 777 m crest length, 7.6 m crest width and 30.5 m width at the base).

- Eder Dam (masonry gravity completed in 1914, 48 m high, 393 m crest length, 6 m crest width and 36 m width at the base).

- Sorpe Dam (earth embankment completed in 1935, 61 m high, 640 m crest length).

Möhne and Eder Dams were hit by 2 bombs each and were breached (77 m wide breach for the Möhne Dam and 50 m wide breach for the Eder Dam) in addition to their powerhouse sustaining heavy damage, leading to wide-scale floods and casualties of approximately 1 650 people, in addition to hampering Nazi Germany's industrial capacity for the war effort. Although hit by 2 bombs, Sorpe Dam sustained only minor damages but no breach. Figures 1 and 2 show photos of the breached dams.

The operation was codenamed "Operation Chastise" and its success was a big boost to the morale of the Allies. A specially developed bomb called "bouncing bomb" was used to attack the dams. It relied on a 500 rpm back spin applied prior to launching to a barrel-shaped bomb, about 1.5 m (5 feet) long and 1.2 m (4 feet) diameter, so that it would bounce on the surface of water when launched from a low altitude of about 18 m (60 feet) and thereby avoid protective torpedo nets that had been deployed upstream of the dams. Upon reaching the dam upstream face, the spin would help the bomb to go down in water while remaining near the dam face and explode at depth by means of a hydrostatic fuse.

More details on these operations, methods used, aircraft, and subsequent damages can be found easily online, including in the list of references below. The operation was also the subject of a 1955 movie called "Dam Busters" which was very popular at that time.

Figure 1
Breached Möhne Dam (Source: Bundesarchiv Bild 101I-637–4192-20, Zerstörte Möhnetalsperre)

Figure 2
Breached Eder Dam (Source: Bundesarchiv Bild 183-C0212–0043-012, Edertalsperre, Zerstörung)

DNIEPROSTROI DAM

Dnieprostroi (Dnjeprostroj) Dam is a 60 m high concrete gravity hydropower dam with a 800 m crest length, completed in 1932 by the Soviet Union on the Dnieper River in Ukraine.

In September 1941, retreating Soviet troops reportedly detonated 90 tons of dynamite in a gallery of the dam. The upper part of the dam, about 200 m long was breached (Figure 3) and the resultant outflow of water caused casualties of between 20 000 and 100 000 people. The exact number is still discussed by historians, as little information is publicly available on the true sequence of events and its consequences. Some of the available online references are listed below. More references in Russian language can be found online. The pressure of the blast caused also heavy damages in the powerhouse. Additionally, retreating German troops dynamited the dam in 1943.

Figure 3
Breached Dnieprostroi Dam (Moroz & Bigg: 2013)

LESSONS LEARNT

To those who wonder if dams are invincible, the answer is unfortunately no. Dams can be damaged or destroyed by natural forces (internal erosion, uplift forces, overtopping, slope failure, etc...), but can also be damaged by man created forces. In both cases, consequences can be extremely damaging.

While the dam demolition actions reported in this case history refer to actions that have taken place during World War II, and have caused very high casualties, it should be noted that hope exists that such dramatic events will never occur again. Namely, the "Protocol Additional to the Geneva Conventions of 12 August 1949, and relating to the Protection of Victims of International Armed Conflicts" dated 8 June 1977 states the following in Article 56 entitled "Protection of works and installations containing dangerous forces":

179

"Works or installations containing dangerous forces, namely dams, dykes and nuclear electrical generating stations, shall not be made the object of attack, even where these objects are military objectives, if such attack may cause the release of dangerous forces and consequent severe losses among the civilian population."

SELECTED REFERENCES

1- Möhne, Eder and Sorpe Dams:

https://en.wikipedia.org/wiki/Operation_Chastise

http://www.raf.mod.uk/history/bombercommanddambusters21march1943.cfm

http://www.thedambusters.org.uk/index.html

2- Dnieprostoi Dam:

https://en.wikipedia.org/wiki/Dnieper_Hydroelectric_Station#WWII_and_Post-war_reconstruction

MOROZ, D. AND BIGG, C., August 23, 2013. *Ukrainian Activists Draw Attention To Little-Known WWII Tragedy*,. http://www.rferl.org/content/european-remembrance-day-ukraine-little-known-ww2-tragedy/25083847.html

3- Other:

JANSEN, R. B., 1980 (reprinted 1983). *Dams and Public Safety*, U.S. Department of the Interior, Bureau of Reclamation, 1980 332 p., Part IV, Significant Accidents and Failures, Dnjeprojstoj dam, p. 135; Eder dam, p.137; Möhne dam, p. 164.

Protocol Additional to the Geneva Conventions of 12 August 1949, and relating to the Protection of Victims of International Armed Conflicts (Protocol I), 8 June 1977, https://www.icrc.org/ihl/INTRO/470

FOLSOM DAM SPILLWAY GATE FAILURE

Pierre Choquet, Canada

ABSTRACT Case history category: c, d, e. This article collects publicly available information about the Folsom Dam spillway gate failure that occurred in 1995 and the subsequent re-evaluation programs on spillway gates that were conducted all across the United States in the subsequent years and led to improved design guidelines.

FOLSOM DAM

Folsom Dam is a concrete gravity dam located approximately 20 km northeast of the city of Sacramento, California on the American River. It was completed in 1956 by the U.S. Corps of engineers and is now operated by the U.S. Bureau of Reclamation and as part of the Central Valley Project . The dam consists in a central concrete gravity dam 104 m height and crest length of 427 m and two embankment wing dams of 44 m height and 640 m crest length for the left wing dam and 44 m height and 2040 m length for the right wing dam (Figure 1).

Figure 1
Folsom Dam and the Folsom Lake (Source: United States Army Corps of Engineers, Michael Nevins - U.S. Army Corps of Engineers photo ID 040316-A-3200N-112)

The spillway of the dam is located at its center and is of the gated overflow type. It is divided in eight equal sections separated by piers. Flow through the spillway is controlled by five service spillway Tainter gates (also called radial gates) and three emergency Tainter gates. The maximum discharge capacity is 14,940 m^3/sec. Outlet works sluiceways consisting of eight 1.5 m x 2.75 m rectangular conduits are located in two tiers through the bottom part of the service spillway.

The Bureau of Reclamation later designed and constructed a power plant with three Francis units for a total capacity of 198 MW. The three power penstocks, visible on Figure 1, pass through monoliths on the left-hand side of the service spillway. Their intakes are one the upstream dam face.

SPILLWAY GATE FAILURE

On the morning of July 17, 1995, the Folsom Dam power plant was shut down and Spillway Gate 3 was opened to maintain flows in the American River. The failure occurred while the gate was operated (Figure 2). There was no warning of structural distress prior to the failure. No one was injured, even though there was a sustained release of 1,132 m³/s into the Lower American River.

Nearly 40 percent of Folsom Lake drained out past the broken gate before it could be repaired. Luckily, no major flooding occurred as a result of the failure, and the Folsom Dam was fully repaired.

Figure 2
Failed Tainter gate No. 3 at Folsom Dam
(Source http://www.pbs.org/wgbh/buildingbig/wonder/structure/folsom.html)

OTHER EVENTS SUBSEQUENT TO GATE FAILURE

A large flood occurred while the damaged gate was being replaced. This required one of the emergency gates to be operated to provide the discharge capacity lost by the damaged gate. Operation of the emergency gate resulted in damage to the impact area downstream of the flip bucket.

Additionally, as a result of the large flood and the ongoing repairs to the spillway gates, the outlet works sluiceways were used more extensively to pass flows. This required all outlets to be used concurrently for an extended period of time, which was not a normal operation. At the conclusion of the extended operation, severe cavitation damage was found in two of the outlet works conduits, and minor cavitation damage was found in other conduits. All these damages were subsequently investigated and repaired (Boyer et al. 2013).

In early 2000s, a comprehensive dam safety review of the Folsom Project was initiated. This review resulted in the replacement of the spillway gate and reinforcement of the spillway piers (Boyer et al. 2013).

INVESTIGATION OF THE SPILLWAY GATE FAILURE

A multi-disciplinary, multi-agency forensic team was immediately formed in 1995 to investigate the spillway gate failure. All of the remaining gates were thoroughly inspected for signs of structural degradation, apart from some corrosion, nothing detrimental was found. The failed gate was removed and then thoroughly examined to determine the mode of failure (Todd, 1999). This examination determined that a diagonal brace joint, adjacent to the trunnion was the initial point of failure. Theoretical finite element models substantiated this study when a trunnion friction coefficient of 0.25 was considered when simulating the forces due to gate lifting. This value had been confirmed also from friction tests performed on the actual trunnion from the failed gate. It was therefore concluded that trunnion friction moment was the key factor, and it had been omitted in the original design calculations.

As a general working principle, the hydraulic load on Tainter gates is transmitted from the cylindrical skin plate, which is in contact with the reservoir, through a number of struts to a convergence at the trunnion hub. The hub collects the load from the struts and transfers it across an interface to the trunnion pin, which is stationary and is connected to the dam. When the gate is operated, the hub rotates around the pin. The struts are primarily compression members, but friction at the pin-hub interface induces a bending stress during gate operation. Typically, and in the case of Folsom Dam as well, the struts are oriented such that the trunnion friction stress is applied to the weak axis of the struts. In order to better handle these loads, the struts are connected with diagonal braces that take the stress as axial loads.

At Folsom Dam, the failure initiated at a diagonal brace between the lowest and second lowest struts. Increasing corrosion at the pin-hub interface had raised the coefficient of friction and, therefore, the bending stress in the strut and the axial force in the brace. The capacity of the brace connection was exceeded, and it failed. This caused the load to redistribute, and the failure progressed, eventually buckling the struts.

A comprehensive re-evaluation program of spillway Tainter gates was initiated in August 1995 by the California Division of Safety of Dams and was conducted on 239 gates on 57 different California dams and involved physical inspection of each gate, finite element modeling of each design, as well as determination of proper loading and acceptance criteria (Schultz et al., 2007). The Federal Energy Regulatory Agency (FERC), in coordination with state dam safety official, launched a similar re-evaluation effort on a nation-wide scale. The U.S. Army Corps of Engineers (USACE) and the U.S. Bureau of reclamation (USBR) also embarked on similar programs to evaluate Tainter gates within their dam inventories.

LESSONS LEARNT

The main lesson learnt is that a coefficient of friction must be considered at the interface of the trunnion hub and trunnion pin in the structural analysis of the lifting of a spillway gate. The generally accepted value, after considerable debate, examination of several trunnion pins and based on a testing program involving strain gauges and lasers is 0.3 and has been adopted by most U.S. agencies involved in Tainter gate analysis (Schultz et al., 2007)

The various re-evaluation programs that were initiated after the Folsom Dam failure lead to a much better understanding of their structural behavior and was at the source of a number of technical articles and design guidelines (Todd, 1999; Schultz et al., 2007; U.S. Army Corps of Engineers, 2000; U.S. Bureau of Reclamation, 1996).

The various studies conducted as part of the re-evaluation program helped also to emphasize the importance of taking into account hydrodynamic loads in the design process, both flow-induced and seismic, and led to technical articles such as U.S. Bureau of Reclamation (2011)

SELECTED REFERENCES

https://en.wikipedia.org/wiki/Folsom_Dam

Modification of Existing Dams-Concrete-Hydrologic, Folsom Dam. In Achievements and Advancements in U.S. Dam Engineering, Edited by Douglas D. Boyer, Richard L. Wiltshire, Glenn S. Tarbox, U.S. Society on Dams, 2013, 792 p.

TODD, R., 1999. *Spillway Tainter Gate Failure at Folsom Dam, California*. In Waterpower '99: Hydro's Future: Technology, Markets, and Policy. Edited by Peggy A. Brookshier

TODD, R., 2002. Determining Earthquake Loading on Spillway gates, Hydro Review,

SCHULTZ, M., JONES, S. AND HUYNH, P., 2007. *Summary of Results from the California Tainter Gate Reevaluation Program*. In Proceedings 27th Annual USSD Conference, Philadelphia, PA, pp. 459–468

U.S. Army Corps of Engineers, 2000. EM 1110–2-2702, *Design of Spillway Tainter gates*.

U.S. Bureau of Reclamation, 1996. Forensic Report on Spillway Gate 3 Failure, Folsom Dam.

U.S. Bureau of Reclamation, 2011. Seismic Induced Loads on Spillway Gates , Phase I - Litterature review. Report DSO-11–06

U.S. Bureau of Reclamation, 2014. Design Standards No. 14: Appurtenant Structures for Dams (Spillways and Outlet Works), Chapter 3: *General Spillway Design Considerations*.

CAHORA BASSA DAM MONITORING

Ilidio Tembe and Chris Oosthuizen, Mozambique

ABSTRACT Case history category: a & e. The chain of monitoring aspects from conception onwards as it has unfolded at Cahora Bassa Dam during the past 40 years is briefly described. Their attention to these aspects, namely, design, installation, observation, maintenance, upgrading, observations, data management and evaluation constitute best monitoring practice for large arch dams.

TECHNICAL DETAILS

The Cahora Bassa Dam, located downstream of Kariba Dam in the Zambezi River, is the largest hydroelectric scheme in Southern Africa. The basic statistics of the double curvature concrete arch dam are as follows: 171 m high, 303 m crest length, 23 m thickness at the base, 4 m thickness at the top, 8 mid-level radial spillway gates and 1 surface gate. The reservoir created by Cahora Bassa dam has a maximum volume of 65×10^9 m^3 (the fourth largest in Africa with a net capacity of 52×10^9 m^3). This massive reservoir, approximately 270 km long and 30 km wide covers an area of 2 900 km^2. The present South bank underground powerhouse has an installed capacity of 5×415 MW and the addition of the planned North bank powerhouse is presently under investigation.

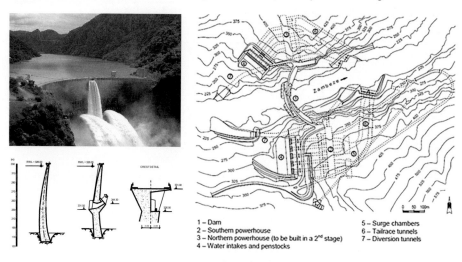

1 – Dam
2 – Southern powerhouse
3 – Northern powerhouse (to be built in a 2nd stage)
4 – Water intakes and penstocks
5 – Surge chambers
6 – Tailrace tunnels
7 – Diversion tunnels

Figure 1
Downstream view, vertical cross sections and layout of Cahora Bassa Dam

Construction commenced in 1969 and impoundment of the dam started in December 1974. The Laboratorio Nacional de Engenharia (LNEC) was directly involved with the design and installation of the monitoring system and with the surveillance of the dam and Power station until 1978. Thereafter the dam was basically operated by HCB staff and dam safety surveillance was mainly carried out by the local HCB staff. LNEC was involved approximately every 5 years with visual dam safety inspections. The original design addressed design assumptions and the ability to monitor the structure's behaviour in the long term. During construction special care has been taken by the instrumentation crew to install the instruments by an experienced team from Portugal assisted by dedicated local personnel.

SURVEILLANCE DETAILS

A concise overview of the instrumentation is given in Table 1. The upgrading from time to time addressed issues that were not foreseen during the original design, such as swelling of the concrete. Only a few instruments stopped functioning after 40 years.

table 1
Details of the monitoring system (parameters, measuring interval & recoding method)

Monitoring equipment	Position			Installation		Number & interval			Parameter			How	
	Dam & found	Powerhouse	Environment	Original	Upgrades	Number	Measuring points	Frequency	Loads	Response	Integrity	Manual	Automated
Water Level													
Gauge Plates		X	X		2	2	2	d			X		
Recorders			X	X		2	3	2w	X			X	
Limnometers			X		X	1	1	2w	X			X	
Climate													
Air temperature			X			2	4	d	X			X	
Relative humidity			X			2	2	d	X			X	
Rain gauge			X			1	1	d	X			X	
Temperature													
Electrical thermometers	X	X		X	X	174	174	2m	X			X	
Electrical strain gauges	X			X		6	12	2m		X	X	X	
Electrical stress gauges	X			X		11	22	2m		X	X	X	
Electrical creep gauges	X			X		4	8	2m		X	X	X	
Electrical joint meters	X			X		5	10	2m		X		X	
Porewater pressures													
Piezometers	X			X	X	62	62	2m		X		X	
Mechanical pore pressure	X			X		6	12	2w		X		X	
Seepage	X	X		X		664	664	m		X		X	

(Continued)

Monitoring equipment	Position			Installation		Number & interval			Parameter			How	
	Dam & found	Powerhouse	Environment	Original	Upgrades	Number	Measuring points	Frequency	Loads	Response	Integrity	Manual	Automated
Geodetic surveys													
Triangulation	X			X	X	49	147	6m		X		X	
Levelling	X	X		X	X	33	33	6m		X		X	
Traversing	X			X	X	2	40	-		X		X	
Vibrations													
Accelerometers	X				X	13	19	c		X	X		X
Seismometers	X	X			X	5	15	c	X		X		X
Displacements													
Pendulums	X			X		7	28	2w		X		X	
Manual extensometers	X	X			X	64	64	2w		X		X	
Electrical extensometers	X	X		X		314	314	2m		X		X	
Electrical joint meters	X			X		5	10	2m		X		X	
Mechanical deformeters	X	X		X	X	184	552	3m		X		X	
Mechanical convergence meters	X	X			X	100	100	3m		X		X	
Deformation													
Electrical strain gauges	X			X		6	12	2m		X	X	X	
Electrical stress gauges	X			X		11	22	2m		X	X	X	
Electrical creep gauges	X			X		4	8	2m		X	X	X	
Other													
Anchorage	X	X				12	20	3m		X		X	

d – daily, w – weekly, m – monthly, c - continuous

DESCRIPTION OF THE INCIDENT

During the mid-1990's the swelling effect of AAR became noticeable and was studied at length for several years by LNEC. It must be noted that the dam was designed and constructed prior to the awareness in Southern Africa of the effect of Alkali Aggregate Reaction (AAR) on dams.

The original monitoring system has been meticulously designed and installed by the designers (LNEC). It was well maintained by trained local personnel but had to be upgraded to address the swelling issue that was not foreseen during the original design. Instruments were therefore required to monitor the rate of swelling. Rod extensometers were installed along the crown cantilever during the late 1990's.

The observed strains and strain rates in the dam wall were carefully assessed and came to the following conclusions as far as the strain rates are concerned (Manitoba Hydro International, 2013):

- The rod extensometer strain rates correlate extremely well with the vertical strain rate results of the originally installed Carlson strain gauges.

- The annual strain rates observed by the convergence meters (between 35–40 and 60 micro strain) are similar to the estimated strain rates of the gate trunnions (around 40 micro strain); and

- The precision levelling results recorded annual vertical swelling strain rates at the crest of between 35 and 40 micro strain at the left flank that decreases to around 22 micro strain on the right flank of the dam wall.

- From the results of the Carlson strain gauges the following is evident:

  - When analysing the average "no stress" strain rates it is evident that the average "stress free" swelling strain rates across the dam is approximately 25 micro strain (see Table 2).

  - The effect of confinement/stress is also evident through the dam section with the average radial strain rates of around 25 micro strain close to both and downstream and upstream faces decreasing to a round only 9 micro strain at the centre as well as decrease of the "no stress" strain rates from 25 to 16 micro strain in the centre of the section.

  - It is important to note that in general there is no significant change in the vertical strain rates through the section of the dam.

  - The apparent "stress free" strain rates as expected decreases with depth to around 20 micro strain at level RL 203 m.

table 2
Carlson strain gauges: Average strain rates

Strain direction (instrument number)	Average strain rate (micro strain/year)		
	1 m from downstream face	Centre of block	1 m from upstream face
Vertical	16.7	17.0	15.6
Upstream/downstream (Radial)	25.8	9.33	26.5
"No stress"	24.0	16.5	25.7

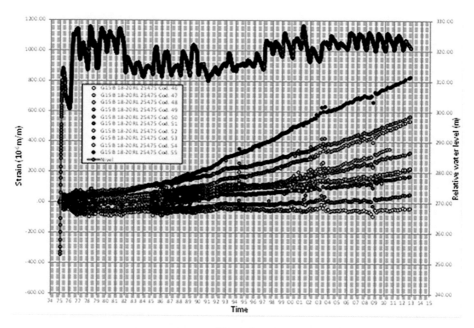

Figure 2
Carlson strain gauges: Strain: RL 254.75 m: Block 18–20: Centre: SACODA results

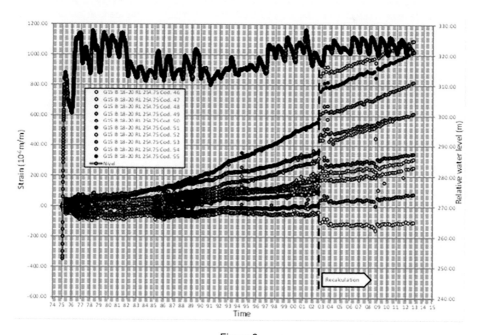

Figure 3
Carlson strain gauges: Strain: RL 254.75 m: Block 18–20: Centre: Recalculation of results using
original calibration constants since 2002

The static monitoring system was enhanced with a dynamic monitoring system in 2010 to measure the modal parameters. Ambient excitation forces such as the natural vibrations of the hydropower plant, outlets, wind, and waves. These systems are usually collectively referred to as Ambient Vibration Monitoring (AVM). The relative positions of the AVM sensors as well as the seismographs are shown in Figure 4.

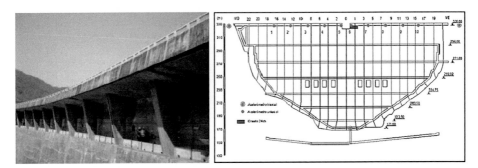

Figure 4
Downstream view and elevation showing the relative positions of the continuously monitored AVM system installed at Cahora Bassa Dam

The AVM (Ambient Vibration Monitoring) system monitors the vibrations near the crest of the wall continuously. The system was upgraded when the internal batteries of the existing seismometers had to be replaced in 2013. The 3 additional tri-axial accelerometers installed in the dam foundation (shown as larger red dots in Figure 4). These accelerometers give the system the capability not only to monitor the dam's response but would also provide the 'reference' seismic loads at the existing dam prior to the envisaged future construction of the left bank power station.

The data collected from the accelerometers is recorded by means of a local data logger at the dam. The data is saved in a file and at hourly intervals sent the central computer located in the dam safety office using the so-called File Transfer Protocol (ftp) via a fibre optic cable. These files are generated in ASCII format (*.txt) in the acquisition system and are processed, managed and stored in binary format (*.bin) in the central computer using the SACODA data management software developed for arch dams (Tecnobar, 2010). The hourly readings are converted to adjusted physical units of acceleration (g). The Frequency Domain Decomposition (FDD) algorithm is used to identify the natural frequencies and related modals parameters, i.e. modal shape and damping. Finally, the software stores these hourly results in binary (*.bin) files. Typical results are presented in Figure 5. Mode shapes during relatively low water levels on 2016-12-26 at 04:00 using FSS method. The dynamic results provide useful information not only to detect minute changes in the behavior of the dam but also for the dynamic calibration of the static Finite Element Models.

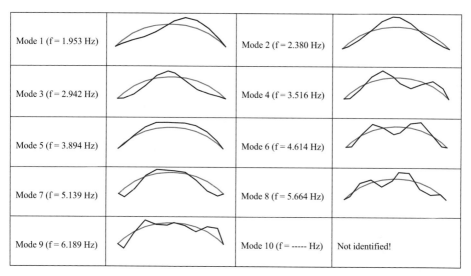

Figure 5
Mode shapes during relatively low water levels on 2016-12-26 at 04:00 using FDD method

LESSONS LEARNT

Dynamic monitoring is a valued technique to monitor changes in the behaviour of arch dams subjected to swelling and Continuous dynamic monitoring (AVM) is considered the ultimate real-time monitoring of the response of arch dams. The state-of-the-art system operational at Cahora Bassa Dam is proof of the high premium placed on dam safety monitoring.

Value of a well –designed, -installed, -maintained, -observed, -analysed and evaluated monitoring system has been underlined. Monitoring is an interdependent chain of activities. If any link of the dam monitoring chain is missing or weak, the whole chain is affected. The phenomenal quality of the results of the old Carlston equipment in the studying the behaviour of the dam has been clearly demonstrates the statement.

Value of diligent dam safety monitoring/surveillance personnel (especially instrumentation technicians). The present dam safety surveillance staff are making a career of dam monitoring and with the necessary management support their diligence is clearly reflected at Cahora Bassa Dam since construction. Skills cannot be developed by training, only through practical experience. The dam safety surveillance team is directly involved with all aspects at Cahora Bassa Dam, from the design, manufacture, and installation of improvements up to the evaluation of the results.

Value of institutional Memory. Institutional memory is one of those less valued aspects of dam safety surveillance. Dedicated personnel making a career of dam safety surveillance are precious sources of institutional memory. A member of the dam safety surveillance team that was part of the original dam instrumentation team at Cahora Bassa Dam, retired in 2015.

SELECTED REFERENCES

Carvalho, E.F. and Tembe, I. (2012): *On-Line Dynamic Monitoring of Cahora Bassa Dam*, published as part of the Proceedings of the Symposium titled "Dams for a Changing Worlds" held during the ICOLD 2012 Congress in Kyoto, Japan

Carvalho EF, Masinge BT & Oosthuizen C (2016). *Monitoring system of Cahora Bassa Dam.... the past, present and way forward.* Proceedings of the ICOLD 2016 Symposium titled "Appropriate Technology to Ensure Proper Development, Operation and Maintenance of Dams in Developing Countries", Sandton, South Africa.

Laboratorio Nacional de Engenharia (1978) Definicao dos sistemas de observacao e constants characteristicas da aparelhagem. Proc. 43/2/3637. Lisbon, Portugal

Laboratório Nacional de Engenharia Civil (2009). *Cahora Bassa Hydroelectric Scheme: Dam behaviour analysis, interpretation and prevision.* Relatorio 341/2009. Lisboa, Portugal.

Manitoba Hydro International (2013) *Cahora Bassa Dam: Evaluation of Monitoring System and Assessment of Dam Performance,* Report authored by LC Hattingh & C Oosthuizen. May 2013. Songo. Mozambique.

TECNOBAR (2010). *Guide to SACODA package – dynamic analysis module.* Lisbon, August 2010. (Portuguese).

ZOEKNOG DAM FAILURE

Chris Oosthuizen, South Africa

ABSTRACT Case history category: e. The installation of the instrumentation system of Zoeknog Dam was done meticulously by an experienced instrumentation expert. He monitored and evaluated the results during first filling (although it was not required from him). His alarm of a potential dam failure was ignored and the dam failed within 2 weeks of his first alarm.

TECHNICAL DETAILS

Zoeknog Dam is located in the North Western corner of the Mpumalanga Province in South Africa. The dam comprises an embankment with morning glory spillway and associated outlet works. It was constructed primarily for irrigation of coffee plantations (for the Lebowa Homeland at the time). The concrete works were constructed by a private contractor and the embankment by the Lebowa government (a construction team with mainly road construction experience). The design and supervision were done by a private consultant. As it was at the time a former "independent homeland" dam it strictly speaking did not fall under the jurisdiction of the South African Dam Safety Regulations. However, the owner undertook to comply with the regulations but did not submit all the necessary documentation required timeously to the Dam Safety Office.

Construction started during 1990 on the 40 m high dam (above lowest foundation level). It was designed as a zoned earthfill section with a central clay core and outer fill zones. The upstream slope varies from 1:4 at the bottom to 1:2,5 at the top. The downstream slope varies from 1:2,5 at the bottom to 1:2 at the top. The central clay core had a slope of 0,8:1 and the 600 mm wide geotextile and sand chimney drain was built in steps. The blanket drain was specified as a "Bidim U44" layer at the bottom with a with a layer of sand and 300mm of gravel and a "Bidim U34" layer at the top. However, the blanket drains as constructed comprised of coarse aggregates up 38mm surround by geotextile.

The embankment material comprised of leached granite from the basin area. Dispersive tests done prior to and during construction tested non-dispersive. Apparently, tests were always done on the upper leached layers and not on the deeper lying material. The deeper layers of the borrowed material was, however, highly dispersive. ASHTO specifications for optimum moisture content (OMC) were used and not Proctor values. The dam was therefore constructed with highly dispersive material compacted at dry of optimum moisture content (a fatal combination).

Figure 1
Zoeknog Dam started impounding December 1992 and failed in January 1993

SURVEILLANCE DETAILS

The layout of the monitoring system (Figure 2) was designed to study the behaviour of the right flank of the embankment comprised the following instruments:

- Piezometers installed at two chainages (460 and 540) with three levels of instruments in each line to measure the phreatic line from upstream of the clay core to downstream of the chimney drain. Both lines were on the right bank, the first 30 metres from the outlet conduit and the second line of instruments 80 metres further

- Three settlement cells were added to the second line to measure the settlement at foundation level.

The installation was done by an experienced private contractor (ex-Department of Water and Sanitation (DWS), Fil Filmalter (Kop-Kop). With the installation of the first row of instruments, he discovered that the blanket drain was not on actual founding level. The contractor apparently planned to install it on the excavation level drawn on the drawings and not on the actual excavated level. Based on his experience at earthfill dams Filmalter apparently complained regularly about the OMC of the clay material as he had to use makeshift "dust masks" during installations when they were placing earthfill material.

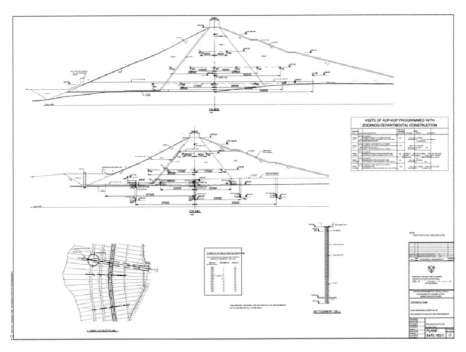

Figure 2
Zoeknog Dam instrumentation drawing

DESCRIPTION OF THE INCIDENT

The events before and after the incident can be summarised as follows:

- Impoundment started just before the December 1992 construction break.

- When the works resumed early January, the instrumentation technician realised on 11 January 1993 that the pressure recorded by the piezometer upstream of the core at chainage 540 m has shot up to the water level in the reservoir.

- The site staff treated it as a false alarm as the earthfill according to them could not be that permeable.

- On 12 January Fil Filmalter was in the author's office. The author told him that the Dam Safety Surveillance teams of DWS (headed by the author) is unfamiliar with the dam as it does not fall under his jurisdiction and referred him to the Dam Safety Office (the dam safety regulating authority) that was indirectly involved. Fil stepped into the passage and shouted in Afrikaans the equivalent of "I am telling you Zoeknog Dam is going to fail and you don't want to do anything" and left.

- On Sunday evening 24 January 1993 around midnight the security guard heard water flowing and cautiously investigated. He found that the flow was not through the outlet pipe but on the left-hand side of the outlets. Six hours later the dam was breached.

- Early morning of 25 January the dam was breached with no lives lost and fortunately no significant damage to other infrastructure.

- Dam Safety Surveillance teams of DWS visited the dam on 2 and 4 February to investigate the failure. It was suspected that it is a piping failure along the conduit and/ or conduit failure leading to the piping failure.

- A week later a taxi driver gave a photo (that he had taken when he passed by early on the morning of the failure) to the site supervising staff. A photo that narrowed the forensic engineers' task.

- The DWS team returned for another full-scale investigation (Barker et al 1993).

- The investigation team did their investigations and finally lodged their report to the Office of the State Attorney on 25 May 1995. As far as this case history is concerned the following remark is of importance "*No blame for the dam failure can be directed at Kop-Kop Instrumentation*" and '*The failure....to carry out regular dam inspections during initial filling of the dam may also have contributed to the dam failure, as signs of distress or seepage could possibly have been detected at an early stage during such inspections, prior to the failure.*" (Keller & van Schalkwyk, 1995).

- The dam was abandoned after the slopes of the breach have been stabilised.

Figure 3
Zoeknog Dam on 2 February 1993 on the left (photo courtesy of Willie Croucamp) and the photo that narrowed the investigation on the right (photo courtesy of an unknown taxi driver)

LESSONS LEARNT

The main lesson learned is to act on warnings of diligent instrumentation personnel. In the case of Zoeknog Dam it would not have mattered. The obvious solution would have been to drain and breach the dam, remove all the earthfill and start again from foundation level.

Value of appropriately- and well experienced instrumentation staff. That is basically a prerequisite for the installation of a well-installed monitoring system.

Figure 4
Abandoned Zoeknog Dam -note the clear signs of dispersive soils

.... In hindsight several design lessons could be learned...:

Dispersiveness of in situ weathered Granitic soils <u>increase</u> with depth (from non-dispersive to highly dispersive). Only surface materials have been tested prior to construction and not thereafter.

Use Procter and not AASHTO compaction specifications for earth-fill embankment dams. The designers used the latter specifications resulting in a less pliable clay core.

Backfill trench excavations with concrete instead of hand tamped clay. (excavation next to the outlet conduit at Zoeknog Dam).

Ensure that the blanket drain is constructed on founding level and not on "assumed foundation levels" indicated on design drawings (the actual levels were up to 5 metres lower in this case).

Use natural sand filters for chimney and blanket drains (not concrete aggregate wrapped in geotextiles as was the case at Zoeknog).

SELECTED REFERENCES

KELLER, H AND VAN SCHALKWYK, A. 1995. Failure of Zoeknog Dam: *Investigation concerning the possible liability of parties involved in the design, construction and supervision of the project.* Report No 1561/ K0001.Office of the State Attorney. Pretoria, South Africa.

BARKER, MB. OOSTHUIZEN,C AND ELGES, HFWK 1993. *Zoeknog dam: Inspection following breaching of the dam which occurred on 25th January 1993.* Department of Water Affairs and Forestry (presently DWS). Pretoria. South Africa.

TOUS DAM FAILURE DUE TO OVERTOPPING

J. Fleitz, Spain

ABSTRACT Case history category: e. The failure of Tous Dam due to overtopping during an extreme flood event in October 1982 originated a complete review of the Spanish dam safety regulations. Many lessons were learnt from the failure: reliable and redundant energy sources to operate electromechanical equipment under extreme conditions, operational flood management based on early warning systems with real-time information, emergency action plans for downstream areas and comprehensive operation manuals including extraordinary and emergency situations have become a standard.

TECHNICAL DETAILS

Tous Dam is located in the lower part of the Jucar River Basin about 15 km from the Mediterranean Sea. It was designed as a multi-purpose reservoir for flood protection and river regulation for water supply and irrigation. The first works to build a dam in Tous started in October 1958, with a project of an 80 m high concrete dam. During construction, the geological conditions of the foundation forced to paralyze the works in December 1964, after location of two faults in the riverbed. The works were resumed in April 1974, modifying the original design into a rock-fill dam with clay core between the remaining concrete blocks previously built at both sides of the riverbed. The modified design was the first stage of a project with a dam crest elevation at 98.5 m to be heightened in a future second stage. The dam structure included a large spillway in the central part provided with 3 radial gates. The total spillway discharge capacity was 7,000 m³/s, corresponding to the 500-year flood as required by the current legislation. The bottom outlet had a capacity of 250 m³/s. The reservoir capacity was 51.5 million m³ at the normal operation level of 84.0 m and 122 million m³ at the crest elevation of 98.5 m.

The main construction works of stage 1 finalized in March 1978 and the first impoundment began at the same time. In November 1979, the maximum normal operation level (84.00 m) was reached and the dam behaviour until the failure in October 1982 was satisfactory without any major incidents. Figure 1 shows the Tous Dam in operation before the failure.

Figure 1
Tous Dam before the failure

DESCRIPTION OF THE INCIDENT

During 20 and 21 October 1982 a particular meteorological condition consisting of a cold, high-altitude depression surrounded by warm air with high moisture content led to extremely heavy rainfall in the hinterland of the central Mediterranean coast of Spain. Average intensities as high as 500 mm in 24 hours were recorded in wide areas. As a result, the Júcar River basin, directly affected by the rains upstream and downstream of Tous Dam, suffered heavy flooding all along. Particularly dramatic was the flooding of the densely populated downstream part of the basin. Tous Dam was the last flood control structure of the Júcar River basin, located only several kilometers upstream of residential areas. It ultimately failed on October 20, at about 19:15 h.

That day the heavy rains quickly filled up the reservoir. The spillway gates were closed and could not be opened as the electric network was out of order due to the weather conditions. Moreover, of the two emergency diesel generators, one was under repair and the other one could not be started. Efforts to raise the gates manually were fruitless.

The overtopping started at 16:50 h (see Figure 2); the maximum water level reached about 1.10 m above the crest at 19:15 h.

Figure 2
Beginning of overtopping of Tous Dam

About 16 hours after recognizing the impossibility of operating the flood gates, the dam was overtopped and washed out after 1 hour by erosion of a greater part of the shoulders and of the central rockfill (see Figure 3). 300 km² of inhabited land, including many towns and villages were severely flooded, affecting around 200 000 people of which 10 000 had to be evacuated. There were 8 casualties, and the damages were estimated to reach $ 400 million, even if part of these damages were likely to be caused by the floods before the arrival of the break wave.

Figure 3
The remains of Tous Dam after the failure

Figure 4 summarizes the reservoir level data which were manually recorded by the responsible dam engineer and the results of the forensic investigations carried out by the *"Centro de Estudios y Experimentación de Obras Públicas"* (CEDEX), a public Spanish civil engineering research agency and the Polytechnic Universities of Madrid and Valencia.

The main conclusions from these investigations are:

- The flood event all over the basin of the Júcar river was extraordinary: In the direct catchment area of Tous (not affected by upstream dams) with an extension of 6,780 km^2 the average precipitation of the event was 277 mm generating a total rainfall volume of 1,880 million m^3. In the downstream part of the basin with an area of 3.692 km^2 the average rainfall was 226 mm (834 million m^3). That means that even if Tous Dam had not failed, there would have been severe flood damages.

- The estimated peak flow of the inflow hydrograph was 10,000 m^3/s equivalent to a 1 000-year flood. In accordance with current regulations, Tous Dam had been designed for a 500-year flood with a peak flow of 7 000 m^3/s.

- The volume of the inflow hydrograph was estimated to be at least 600 million m^3, about 12 times the reservoir volume at normal operation level and 5 times at dam crest level.

- The peak inflow of approx. 10,000 m^3/s was reached after 10 pm on October 20, which is about three hours after the beginning of the dam breach.

- The dam breach wave had a maximum peak flow of 15 000 m^3/s.

- Numerical reservoir routing and physical reduced scale models showed that even with a totally empty reservoir in the beginning of the flood event and fully open gates, the failure probably could not have been avoided. The load scenario of a 1 000-year flood was beyond the design assumptions.

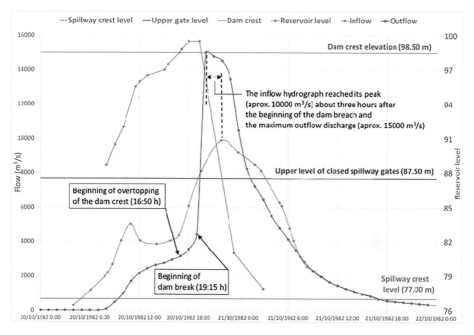

Figure 4
Inflow and outflow hydrographs and reservoir level evolution during October 20 and 21 at Tous Dam
(Source: Témez et al. and Arenillas et al.)

LESSONS LEARNT

In the context of the present bulletin on surveillance case histories, the main lesson to be learnt from the Tous Dam failure is the utmost importance of reliable and redundant energy sources to operate all essential devices and especially the electromechanical equipment of the outlet works, mainly under extreme conditions.

Two engineers of the technical team in charge of the operation of the dam were condemned by the Spanish legal authorities because of gross negligence in the longest litigation in Spanish history of justice. The parallel engineering trial came to a different conclusion: the flood event was far beyond the design criteria correctly adapted according to the current regulations. Even perfectly operating gates or totally open gates would not have provided sufficient routing capacity to avoid the raise of the reservoir beyond the dam crest and to prevent the failure.

The failure of Tous Dam had a very strong impact in the Spanish society and especially in the dam engineering community and among responsible dam engineers. The latter found themselves defenceless and without legal protection.

This situation originated a complete review of the dam safety regulations, enhancing design criteria especially for embankment dams but also focusing on adequate operation procedures, regular safety inspections and emergency action plans. Tous Dam was also the origin for implementing automatic hydro meteorological networks throughout the Spanish River basins to collect real-time data and to improve flood forecast and management. Further it contributed to the international debate about the pros and cons of gated spillways. After the Tous Dam failure the majority of the dams in the Spanish Mediterranean basins changed their operation manuals leaving their gates permanently open.

A new Tous Dam (see Figure 5) was built on the same site and part of the clay core material, which had shown a relatively high resistance to water flow, was reused for constructing the new dam. The main characteristics of the new dam compared to the one that failed (see Table 7), show the impact of the new safety standards:

Figure 5
The new Tous Dam

Table 1
Comparison between the old and the new Tous Dam

	Old dam	New dam
Dam height	46.5 m	110.5 m
Crest elevation and reservoir volume	98.5 m / 122 million m³	162.5 m / 792 million m³
Normal operation level and volume	84.0 m / 51.5 million m³	130.0 m / 379 million m³ (90.0 m / 80 million m³ during flood season)
Spillway capacity	7 000 m³/s (gated)	20 000 m³/s (ungated)

SELECTED REFERENCES

UTRILLAS SERRANO J.L., 2013, La presa de Tous, Ingeniería, seguridad y desarrollo en la Ribera del Júcar, Confederación Hidrográfica del Júcar.

TÉMEZ J.R. & MATEOS C. (CEDEX), 1993, Hidrograma de entrada a Tous. Datos para un juicio crítico, Revista de Obras Públicas N° 3319, Año 140, March 1993

ARENILLAS M., MARTÍNEZ R. (Polytechnic University of Madrid), Cortés R., Ferri J.A. & Botella J. (Polytechnic University of Valencia), Nuevos datos sobre la crecida del Júcar de octubre de 1982, Revista de Obras Públicas N° 3323, Año 140, July-August 1993

ALCRUDO F. & MULET J., 2007, Description of the Tous Dam break case study (Spain), Journal of Hydraulic Research Vol. 45, Iss. sup1,2007

DE WRACHIEN D. & MAMBRETTI S., 2009, Dam-break Problems, Solutions and Case Studies edited by D. de Wrachien (State University of Milan, Italy) and S. Mambretti (Politecnico di Milano, Italy), WIT PRESS 2009

6.5   ONE PAGE SUMMARIES OF CASE HISTORIES

EL CHÓCON DAM - INTERNAL EROSION

Dam type: Embankment - 86m high
Description: Zoned earthfill with central core
Case history category: e, c, d
Main objective: Early detection of failure mechanisms
Main benefit: Correct and on time remedial action
Observations: Early detection of potential internal erosion and assessment of remedial works

A continuous increase of piezometric levels in the rock-core contact on right abutment was observed, leading to the design of the following remedial works: a grouting program considered necessary at both the right and left banks as well as for the entire foundation in order to reduce the potential internal erosion of the clay core into open foundation joints.

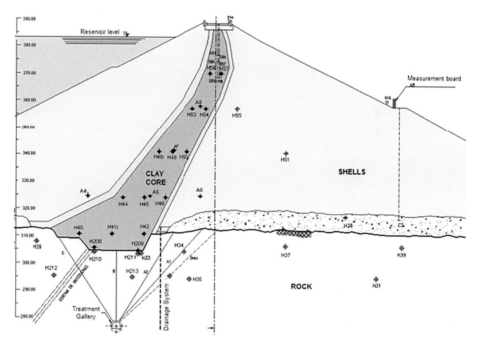

Cross section of dam

Thus, the objective of the remedial works was to reduce the seepage through the open foundation joints, caused either by gypsum dissolution or as a result of the original grouting curtain not being completely effective.

DOI: 10.1201/9781003274841-10

LESSONS LEARNT

The importance of having reliable and well-located instruments at the rock contact interface. Piezometric measurements at the rock contact interface facilitated the detection of the increase in piezometric pressures.

Pro-active surveillance of dams during their life cycle became the standard. This problem detected by effective monitoring lead to further investigations and remedial works also highlighting the importance of using with an Independent Panel of Experts.

Proper instrumentation is not just essential for monitoring during the foundation remedial work but also for the future long-term behaviour of the remedial work. The design of the remedial works included an important group of piezometers to check the short as well as long term efficiency of the new grouting and drainage curtains.

DURLASSBODEN DAM

Dam type: Embankment – 80 m high
Description: Zoned earthfill with central clay ore
Case history category: c, d
Main objective: Surveillance of the grout curtain
Main benefit: Good understanding of dam behaviour
Observations: Importance of the long-term behaviour of deep dam foundations

Durlassboden Dam is an embankment dam with a central clay core and is founded on erodible alluvial deposits with a maximum depth of 130 m. Due to economic and technical reasons the grout curtain was not carried out to the full depth of the alluvium, but only down to a silt layer at a depth of about 50 m. Therefore, water can pass through the "window" below the silt layer.

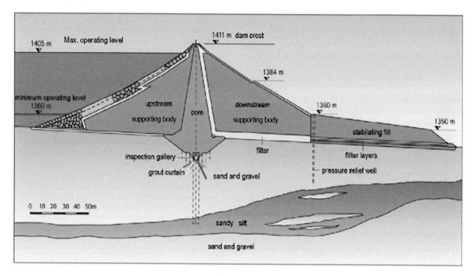

Cross section of the dam

These adverse foundation conditions required an extensive monitoring system that consists of relief wells at the downstream toe, measuring weirs for measuring discharge and turbidity in the area downstream of the dam, piezometers in the area downstream of the embankment and piezometers upstream and downstream of the grout curtain (see coloured lines in the Figure above)

In 2010, after more than 40 years of operation the question was posed whether an upgrade of a part of the area downstream of the dam or even of the grout curtain is necessary. To get the right answer to this question an intensive evaluation of the measured data was carried out. Having data over a period of more than 40 years should provide answers on whether any positive or negative trends could be identified. The evaluation showed clearly that there are no negative trends.

LESSONS LEARNT

The evaluation demonstrated the satisfying long-term performance of the dam and the grout curtain over the decades. Any necessary improvements may be based on current and future readings.

GMUEND DAM

Dam type: Gravity dam – 37 m high
Description: Concrete
Case history category: a
Main objective: Influence of material of instruments on results
Main benefit: Improvement of measurement results
Observations: Totally different measurement results after change of instrument material

Gmuend Dam was initially constructed in 1943–45 as a concrete arch dam. Due to stability problems at the left abutments a gravity dam was added downstream of the arch dam in 1964. A further improvement and slight raising of the dam was carried out in 1991. The dam is located in the Austrian Alps at an elevation of about 1,200 m a.s.l. The annual ambient temperature variation amounts to about 40°C, from -20° to +20° C.

Cross section

View from downstream   Cross section

The left flank downstream of the dam is now stabilized by a retaining wall with 52 anchors and it is monitored, amongst others, with two extensometers with lengths of 12 m and 31.5 m. Both of them were equipped with ordinary steel bars and the readings showed an annual variation of about 1 mm with the minimum of the measured distance in autumn and the maximum in springtime.

To detect the influence of the temperature variation on the measuring system (the extensometer), the steel bar of the 12 m long extensometer was replaced by a bar made of invar in 2006. Invar was selected because it exhibits a coefficient of thermal expansion of about 1/10 of that of steel. This resulted in a significant change of the measurement results.

LESSONS LEARNT

The annual variation of the invar-extensometers is now only about 0.4 mm with the maximum in autumn and the minimum in springtime. Obviously, the measuring instrument is affected by the ambient temperature far more than the subject (the deformation of the rock), which should be measured.

ZILLERGRÜNDL DAM

Dam type: Arch dam
Description: Double curvature concrete
Case history category: a, b, c, d
Main objective: Surveillance of dam foundation
Main benefit: Controlled uplift in the dam foundation
Observations: Good correlation of predicted and measured dam behaviour

In the valley bottom and in the abutments of the 186 m high Zillergründl Dam, the dam is provided with an upstream apron which reaches up to a little more than half the height of the abutments. In addition to this, a flexible movement joint was provided in the valley bottom on the upstream side close to concrete foundation interface in the valley bottom. With appropriate drain pipes the water pressure in the movement joint is limited to elevation 1 745 m a.s.l, and below the base gallery to elevation 1 700 m a.s.l. Fig. A) shows a cross section of the dam and Fig. B) shows the base section in more detail together with the uplift pressure along the base joint.

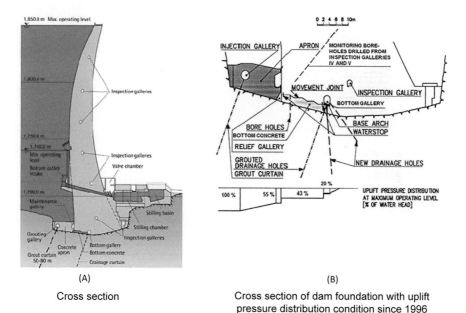

(A)

Cross section

(B)

Cross section of dam foundation with uplift pressure distribution condition since 1996

The Zillergründl Dam, especially the base area of the dam, is comprehensively equipped with instruments, all in all with more than 1 000 measuring devices. This enables a comparison of measured and calculated/predicted behaviour.

LESSONS LEARNT

This case shows that measurements with suitable monitoring equipment at a suitable location can confirm or help to adjust calculation methods. Although predicted that the flexible movement joint would remain in contact even at a full reservoir, the measurement results on the other hand showed that the joint opens 6 mm, as mentioned above. A nonlinear FEM analysis carried out later, showed a dam behaviour in accordance with the results of the measurements. Observations from a seismic event in 2011 also confirmed the validity of the nonlinear model when comparing the observed and calculated modal frequencies.

COMOÉ DAM - EROSION IN THE FOUNDATION

Dam type: Embankment dam - 25m high and 1165m long, founded on lateritic ground
Description: Homogeneous earthfill
Case history category: b, c, e
Main objective: highlight the specific behaviour of tropical soils and the importance of proper interpretation of monitoring results
Main benefit: improving the safety of the dam without emptying the reservoir
Observations: large voids developed by erosion in the dam foundation, where canaliculus were observed during construction

A continuous increase of uncontrolled seepage flow on both flanks, collected downstream of the dam along with subsidence development area downstream of the toe of the left bank, was attributed to internal erosion occurring in the foundation. A deep foundation treatment was carried out from the crest of the dam. It included sheet piling, tube à manchette grouting and compaction grouting.

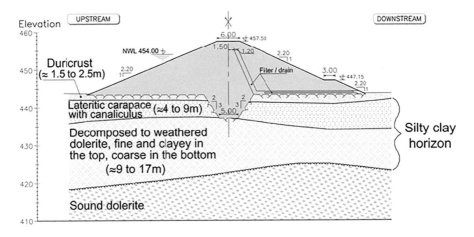

Cross section of dam (see page concentrated in the silty clay horizon)

The objectives of the foundation treatment were (1) to get a positive cut-off in the most vulnerable zones of the foundation and (2) to fill the voids developed in the silty clay horizon under the downstream shell of the embankment, on both flanks. The main indicator of the effectiveness of the treatment is the decrease of the seepage low and downstream hydrostatic pressures. The original treatment was concentrated on the central part of the dam where the foundation behaviour was satisfactory.

LESSONS LEARNT

Lateritic foundation may be subjected to piping even under small hydraulic head. It is now established that under some conditions, "laterite karst" may develop, mainly by erosion.

Proper interpretation with graphical representation of statistical analysis results is very useful in highlighting some parameters particularly for non-initiated owners.

CAMEROUN SONG LOULOU DAM – AAR

Dam type: Embankment with gravity spillway section – 35m high with 8x48 MW hydropower units
Description: Earthfill and concrete
Case history category: c, e
Main objective: Effective monitoring of concrete expansion due to AAR development.
Main benefit: Monitoring system had to be improved to facilitate AAR monitoring.
Observations: AAR could have a significant impact on dam safety and operation depending on the rate and the amplitude of expansion.

At Song Loulou Dam, the mass concrete for the gravity section contained aggregate produced from dam foundation excavation material, corresponding to a strong micaceous gneiss interlayed with amphibolite lenses. D_{max} was 50 mm and OPC content was ranging from 325 to 335 kg/m³. Unconfined compressive strength reached an average of 22.5 MPa.

The first signs of concrete swelling were observed in 1991, almost 12 years after the first concrete placement. Concrete expansion was attributed to both alkali-aggregate reaction (AAR) and internal sulphate attack.

The rate of expansion was estimated to 150 μ strain/year, according to the more reliable geodetic observations at the crest of the intake section. The main consequences of this situation are the following:

- Threat on the safety of the dam:

    – Development of opened horizontal cracks in the intake structure, in direct contact with the reservoir. The increase in the uplift thus generated can lead to the failure.

    – The spillway gates may be blocked. They can no longer be operated, which can lead to the overtopping of the embankment section in case of large flooding.

- Blockage of the turbines with loss in power production; and

- Failure of the spillway pre-stressed tendons supporting the gate thrusts.

The monitoring system installed during construction operated satisfactorily only during the first years. It had to be improved later to get reliable data on the dam deformation, spillway gates and powerhouse behaviour caused by the concrete swelling.

Subsequent to upgrading of the monitoring system as well as further studies and laboratory testing, it was concluded in 2012–2013 that the swelling due to AAR is nearing its end. This made it possible to consider carrying out rehabilitation works to spread out over a period of approximately 10 years at the total cost of € 110 million. It includes further monitoring system improvements.

LESSONS LEARNT

AAR could have a significant impact on dam safety and operation depending on the rate and the amplitude of expansion.

A monitoring system, well designed, properly constructed and maintained in a good condition is of great help in the early detection and understanding of the AAR phenomenon.

DETECTION OF WATERLINE LEAKS IN A CONCRETE FACED ROCKFILL DAM

Dam type: Embankment - 27m high
Description: Concrete face rockfill
Case history category: b, c
Main objective: Detecting sources of leakage at joints between concrete slabs
Main benefit: Target remedial actions
Observations: Long term weir flow measurements

The Kootenay Canal facility is located on the Kootenay River near the city of Castlegar in British Columbia, Canada, and was completed in 1976. The facility consists of a 5 km long canal leading to a 30 m high concrete intake dam with four penstocks. The forebay includes two concrete faced rockfill dams roughly 27 m in height on the north and south side of the canal. The forebay slopes are lined with concrete slabs founded on rockfill. Leakage through the embankments is believed to originate at slab joints. Remedial work including a geomembrane liner and a filtered drain at the downstream toe was completed in 2009, but leakage could still be measured in the weir L4 downstream of the concrete intake dam.

It was postulated that analysis of the correlation between forebay water level and Weir L4 flow could provide guidance on leak elevations. The analysis was completed using an energy balance orifice flow equation: $Q = C_D \cdot A \cdot \sqrt{2 \cdot g \cdot (H - Z)}$

The useful insight gained from the analysis is that only high elevation defects can produce large changes in flow for relatively moderate changes of forebay elevation, as illustrated in the Figure below.

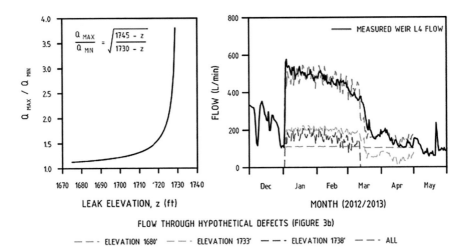

FLOW THROUGH HYPOTHETICAL DEFECTS (FIGURE 3b)

— — · ELEVATION 1680'   — — · ELEVATION 1733'   — — · ELEVATION 1738'   — — · ALL

LESSONS LEARNT

For new construction, rigorous documentation of as-built details for features such as curb walls and conveyance pipes may be of high value for later use by surveillance personnel. For example, at Kootenay Canal uncertainty regarding the terminus of a buried curb wall led to years of speculation about and investigation of the embankment flow regime.

The results of persistent, long-term investigations by surveillance personnel may have unexpected applications. The analysis described herein, which contributed to optimization of the final design of the 2014 geomembrane, would not have been possible if the sequence of events leading to the Weir L4 upgrade had not occurred.

The proportionality of orifice flow to the square of the driving head is such that defects that develop in low head (shallow) locations can contribute significantly to leakage through concrete-faced rockfill embankments.

INSTRUMENT PERFORMANCE EVALUATION USING TRENDS ANALYSIS

Dam type: Embankment - 180m high
Description: Zoned earthfill with central clay core
Case history category: b
Main objective: Model relationship between an instrument and its key driver variable
Main benefit: Improve future dam performance predictions
Observations: Long term weir flow measurements

Trends analysis can be easily implemented in an Excel Spreadsheet to model the relationships between an instrument and its key driver variable. As an example, the relationship between instrument and reservoir level readings can be expressed as $P(t) = A_0 + C_0*(t - t_0) + A_1*R(t - t_g)$, where, $P(t)$ = the instrument reading at time t, $R(t - t_g)$ = the reservoir level reading at time $(t - t_g)$; t_g = the time lag between the instrument and reservoir readings (i.e. travel time of flow from the reservoir to the point where the instrument measured); C_0 = the rate of change (creep) of instrument readings with time; t_0 = a reference time for creep correction; A_0 and A_1 are constants.

WAC Bennett Dam is a 180 m high, 2 km long earthfill dam, located on the Peace River in north eastern British Columbia, Canada. Weir 6, which measures seepage flows from the Right Bank Terrace and downstream of the 1996 Sinkhole, is an important instrument regarding dam performance. The figure below shows recorded and predicted flows of Weir 6 and reservoir level readings since the dam was constructed.

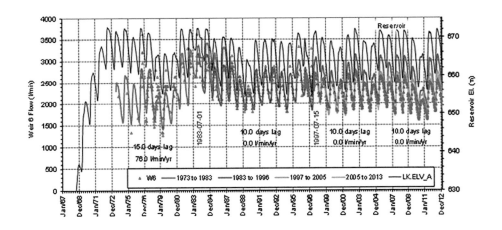

LESSONS LEARNT

Trends analysis is a useful tool to analyse historical instrumentation data and can be used to evaluate the performance of an instrument.

The importance of incorporating lag time and creep in the trends analysis was emphasized. The relationship between a driver and instrument has greatly improved by incorporating the lag time and creep. A narrow and more precise performance bound can be obtained to predict the future dam performance.

Consideration should be given to separate other factors from the correlation equations in a trend's analysis. Other factors such as precipitation, run-off or instrument errors may skew the correlation. A more complex model incorporating other important factors is recommended if necessary.

UNDERWATER INSPECTION OF WAC BENNETT DAM

Dam type: Embankment - 180m high
Description: Zoned earthfill with central clay core
Case history category: c, d
Main objective: Identify and monitor features in the underwater portion of the dam
Main benefit: Create a full 3-dimensional model of the upstream surface of the dam
Observations: Simultaneous multi-beam and side scan baseline surveys

In 2013, BC Hydro undertook a comprehensive remote underwater inspection of the entire underwater surface of the WAC Bennett Dam, located on the Peace River in north-eastern British Columbia, Canada. The dam is a zoned 183 m high earthfill embankment that retains the 360 km long Williston Reservoir.

The objective of the survey was to establish a highly accurate baseline surface to (i) identify and characterise previously unknown features on the dam surface and (ii) to establish performance criteria and methodology for underwater monitoring.

The major work items included the following: a simultaneous multi-beam and side scan baseline survey, on-site multi-beam data analysis and reporting to identify features of interest, high-resolution re-inspection, and characterisation of the features of interest, data integration into 3D GIS and CAD models. The result was a complete, high quality and accurate 3D surface of the underwater portion of the earth fill dam as shown in the Figure.

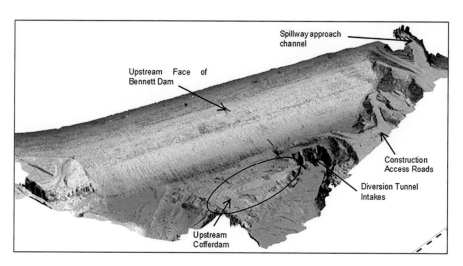

Image of 3D dam surface produced from multibeam data

LESSONS LEARNT

The WAC Bennett Dam underwater survey and inspection was successful at producing data of sufficient quality and accuracy to allow the identification of previously unknown features over the coverage area. This work demonstrated that, for the most part, the industry and underwater technology is at a state at which underwater surveys can be performed to significant water depths with enough accuracy for dam surveillance and monitoring. This provides benefits over historical piecemeal inspections and can be used to provide comprehensive qualitative data over a large area.

When registered into a coherent data set, the data collected using the described underwater ROV collection methods provided an accurate representation of the dam surface. Features within the point cloud data were easily identifiable through the use of point cloud viewing software. The use of more specialized GIS software can improve the ability to assess a feature and compare it with historical and other information. Higher resolution scans using the same multibeam sensor but flying the ROV at lower altitude and at smaller line spacing together with stationary high-resolution sonar scans provided excellent detail of features of interest allowing for accurate characterisation and future monitoring.

HETEROGENEITY OF AN EMBANKMENT DAM CORE

Dam type: Embankment – 94.5 m high
Description: Sand/gravel fill with central till core
Case history category: a, b
Main objective: Assessing a zone of higher hydraulic conductivity in the central till core
Main benefit: Decide if remedial actions were required
Observations: Using construction control data and as-built reports

The surveillance of a 94.5 m high sand and gravel fill embankment dam in northern Quebec, Canada, by means of temperature monitoring revealed the existence of a zone of higher hydraulic conductivity in its central till core.

The presence of a more pervious zone may indicate cracking and/or ongoing internal erosion phenomena which could be detrimental to the safety of the dam. Its presence can also be related to variations in till properties due to construction procedures.

Field tests were done on the compacted till for construction control. These included mainly the determination of grain-size distribution, water content and density. The spatial continuity of the measured fines content in the dam during construction was computed using geostatistics and used to predict values at unsampled locations and for the entire core volume. The hydraulic conductivities were subsequently inferred.

The representativeness of the inferred hydraulic conductivities was then assessed with temperature monitoring data. The figure below shows the annual temperature amplitudes contour lines superposed with the hydraulic conductivity layering in the core.

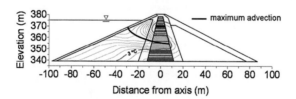

Inferred hydraulic conductivities and line of maximum advection

The heterogeneities and zones of higher hydraulic conductivity in the dam core were found to be mainly due to variations of fines content and till fabric, which depends on water content and density that occurred during construction. No internal erosion or cracking phenomena are present in the core. No other remedial actions are needed.

LESSONS LEARNT

Construction control data and as-built reports are essential for dam safety monitoring data interpretation. This background information provides the historical context for interpretation. The effective synthesis of this information by means of geostatistical analyses can help the dam owner in determining the necessity for remedial works.

The variability of soil properties must be considered in dam safety assessments. The knowledge of the geological deposition mechanisms of materials used in dams and the effects of construction practices on their geotechnical properties can explain field observations and measurements that could be otherwise attributed to detrimental effects such as internal erosion or cracking.

SEEPAGE DETECTION IN AN EMBANKMENT DAM FOUNDATION USING MODERN FIELD AND NUMERICAL TOOLS

Dam type: Embankment dam - 15 m high
Description: Sand/gravel fill with central till core
Case history category: c, e
Main objective: Minimize internal erosion in an underlying sand layer
Main benefit: Design of targeted relief wells as stabilizing measure
Observations: Electromagnetic and Lidar surveys as well as geostatistical analyses

Significant seepage was observed downstream from an embankment dam founded on a clay layer underlain by sand located in northern Québec, Canada. This structure is a 15 m high, and 1 200 m long embankment founded on marine clay underlain by a fine sand deposit of up to 70 m in thickness.

The risk of internal erosion of the sand layer through the overlying clay is increased where the factor of safety against uplift is lower. The calculation of factors of safety against uplift requires an estimation of clay thickness and pore pressures for the entire area. A global electromagnetic survey was realized to detect the main sources, pathways and exits of seepage from the sand layer. An airborne LIDAR survey was performed to determine land surface elevations. Several boreholes and cone penetrations tests were realized over the years to assess the foundation stratigraphy and open-tube piezometers were installed. Clay-sand contact elevations and piezometric levels in the sand layer were estimated using geostatistics. A factor of safety against uplift was then computed for the entire downstream area, as shown in the Figure. The design of stabilizing measures was specifically targeted at the more critical areas where factors of safety was lower. Several relief wells were installed, and their drainage effect caused a decrease of piezometric levels in the sand layer in the lake area. This decrease of pore pressures increased the factor of safety against uplift to acceptable values and thus decreased the likelihood of internal erosion.

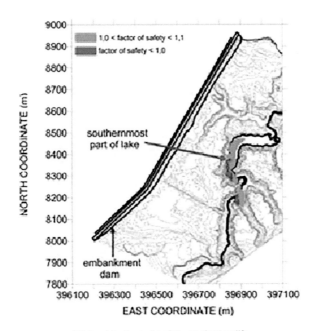

Minimal factors of safety against uplift

LESSONS LEARNT

Regular visual inspections are of utmost importance in the detection of adverse phenomena. Inspections, sometimes performed in difficult conditions, provide the earliest warnings regarding the evolution of dam behaviour. A thorough knowledge of the potential failure mechanisms pertaining to a dam is also required to evaluate the significance of observations.

A dam owner has to be aware of the latest field and numerical tools to complement and synthesize available data. The use of electromagnetic and laser surveys as well as geostatistical analyses were instrumental in defining the limits of an otherwise spread-out problem and to provide an optimal solution, in terms of costs, feasibility and effectiveness, for the long-term stabilization of the foundation.

ASSESSMENT OF THE PORCE II DAM

Dam type: Gravity with embankment on right flank - 122m high
Description: Roller Compacted concrete (RCC) and homogenous earthfill
Case history category: c, d, e
Main objective: Evaluate the safety of the dam using the results of the monitoring system
Main benefit: Detect and timely attend deficiencies in the dam
Observations: The evaluation of the monitoring results facilitate the evaluation of the safety of the dam

After 12 years of operation, EPM decided to carry out an assessment of Porce II Dam to evaluate the dam stability, to verify the design assumptions and to compare the design criteria with the current state of the art. These different evaluations but especially the analysis of the monitoring results allowed for the comparison of the performance of the dam with the original design assumptions and made it possible to identify any component that does not comply with current design standards.

Porce II Dam has a total 278 instruments to measure seepage, pore pressures, deformation, temperature, and acceleration, located in the body of the dam and its foundations and adjacent areas. This case history presents some of the most important aspects obtained from the evaluation of the monitoring results from 2001 to date.

General view of Porce II Dam

From the analysis of the monitoring results and its inputs into numerical stability models it was possible to make conclusions about the safety of the dam including its stability under both static and well as dynamic conditions. In addition, the analysis of the monitoring results allowed to detect changes in the records to give warnings and evaluate if anomalous behaviors were presented; so that they could take timely corrective measures. It was concluded that the behavior of the dam is adequate according to current criteria of dam engineering and its level of safety is satisfactory.

LESSONS LEARNT

The monitoring results were useful to determine the alarm thresholds for either the initiation or the update of the emergency action plan.

Thanks to the assessment study it was possible to determine the need to install extra piezometers on the earthfill dam to get data for a better understanding of the behaviour of this part of the dam.

PORCE DAM (COLOMBIA): METHODOLOGY FOR THE DEFINITION OF MONITORING THRESHOLDS

Dam type: Gravity with embankment on right flank - 122m high
Description: Roller Compacted concrete (RCC) and homogenous earthfill
Case history category: c, e
Main objective: Definition of monitoring thresholds
Main benefit: Diagnose and prevent the development of possible failure modes
Observations: The definition of alert or alarm thresholds is a fundamental basis for identifying emergency situations that may affect the safety of the dam.

Within the procedures for risk management inherent in a civil structure, such as a dam, it is essential to constantly monitor of all the influential variables in the performance of the dam. This will warn of the presence of any potential threat so that corrective measures can be taken in time to avoid the occurrence of a failure mechanism. It important to note that proper interpretation of monitoring information as well as the observations from the visual inspections are essential to properly understand the behavior of a dam. The results are also used to diagnose and prevent the development of possible failure modes when monitoring information exceed a preset alarm level or threshold.

This case history describes the methodology used to define monitoring thresholds and explain why the methodology typically applied in some parts of Europe, where recorded variables are correlated with external variables cannot be applied successfully in tropical countries and therefore it is necessary to propose another methodology to define thresholds directly on the statistics data of the monitoring variables.

The proposed methodology for the analysis of monitoring information, the definition of monitoring thresholds and their possible correlation with the potential failure modes are provided in the following sequence:

- Compilation of monitoring information.

- Debugging information.

- Statistical information processing: Correlating the monitoring variables with the external variables and performing statistical analysis directly on the monitoring variables.

- Definition of monitoring thresholds it includes both quantitative (monitoring results from instrumentation) as well as qualitative (visual inspection) aspects; and

- Correlation of monitoring thresholds with previously identified potential failure modes.

LESSONS LEARNT

An alternative method for the definition of thresholds has been presented, based on statistical analysis of the monitoring information, which departs from the classic analysis of the correlation of the monitoring variables with the external variables, basically due to the low correlation that is typical for tropical countries like Colombia. In its absence, the method makes an explicit separation of random and epistemic uncertainties and proposes the use of direct statistical analyses of the data for the definition of the thresholds.

Once the thresholds have been defined using both quantitative (monitoring results from instrumentation) as well as qualitative (visual inspection) means, these are then correlated with the potential failure modes.

The proper surveillance and subsequent analysis constitute the basic tool to improve the understanding of the potential failure modes.

SANTA RITA DAM (COLOMBIA): DETECTING ANOMALOUS BEHAVIOR OF PORE PRESSURES IN EARTHFILL DAMS AND ASSESMENT STUDY

Dam type: Three embankments – 51.5 m high
Description: Earthfill – silty core at main embankment
Case history category: e, c, d
Main objective: Early detection of failure mechanisms
Main benefit: Evaluate the effectiveness of the rehabilitation or update works
Observations: Reduce of liquefaction potential and assessment of update works

During the regular visual inspections carried out at the dam, erosion on the upstream slope and wet areas downstream of the embankment were observed. It was also observed that the body of the dam was saturated, and the seepage was flowing through the embankment. In the area downstream of the embankment some wet areas were observed. These anomalies led to an assessment study to determinate the safety of Santa Rita Dam.

Upgrades at Auxiliary II embankment of Santa Rita Dam done in 2004

The assessment study concluded that a high liquefaction potential in Auxiliary II embankment that could generate mass movements during the occurrence of an exceptional earthquake. The solutions proposed by the assessment were focussed on the reduction of the liquefaction potential of soil, the control of seepage flow through the left abutment and the reduction of the effect of possible increases in the seepage flow through this abutment during an earthquake. The update works were carry out in 2004. The main element of the upgrade to reduce the liquefaction potential, a group of gravel wells were installed to increase the effective stresses in the most susceptible areas. To protect the left abutment, a blanket drain/filter and a reinforcement fill were added on the downstream side, to control fines migration and ensure efficient drainage system. In addition, geotechnical instrumentation (vibrating wire piezometers and seepage weirs) was installed to improve monitoring of the dam and to verify the effectiveness of the upgrades in the long term.

LESSONS LEARNT

An efficient instrumentation program should: provide early warning about a potential problem, assist in the evaluation of a detected problem, and evaluate the effectiveness of the rehabilitation or upgrades. Despite the initial low number of instruments in the Auxiliary II embankment, the available results still made it possible to detect excessive seepage from the reservoir into the body of the dam, to evaluate the criticality of the problem and to highlight the need improved monitoring. It resulted in the installation of more instruments, which currently continues the permanent monitoring of the area.

It is important to complement the analysis of the instrumentation with the results of visual inspections, to have certainty about the problem evaluated. The visual inspections confirm the monitoring observations as seepage and wet areas were observed in the same places where the instruments recorded anomalies.

TONA DAM (COLOMBIA): FIELD ROCKFILL DEFORMATION MODULUS

Dam type: Embankment - 103 meters high

Description: Concrete faced rockfill (CFRD)

Case history category: a, c

Main objective: Compare the deformation modulus of large-scale laboratory tests and available literature with observed values

Main benefit: The laboratory test generates a good approximation of the magnitude of the rockfill deformation modulus

Observations: A proper estimation of the deformation modulus allows one to optimize the design and lower the risks of the design

The design of concrete face rockfill dams requires the control of cracks in concrete elements, principally on the face and plinth and the resulting increase in seepage flows and loss of volume of fill material. In this case history, the monitoring results from the instrumentation were used to estimate the deformation modulus for comparison with the those estimated from the large-scale tests and those determined using methods used during design described in literature to improve the methodologies to estimate the deformation modulus.

This paper addresses the specific case of the Tona River Dam which is a Concrete Face Rockfill Dam (CFRD) currently under construction for the urban water supply Bucaramanga city. The dam, 103 m high, 248 m long with a crest width of 9 m has upstream and downstream slopes of 1.4H:1.0V and 1.5H:1.0V respectively. The total rockfill volume is about 1,9 million m³. For this case history the process used to determine the rockfill modulus during design are described in more detail. Subsequently, comparisons with estimates from large scale tests and those obtained from the monitoring results from instrumentation installed during the construction of the dam were made.

LESSONS LEARNT

From literature, a wide range of values for the deformation modulus were estimated. Nevertheless, these values are useful for an initial design. It is necessary to refine these values in order to execute a more realistic design in terms of displacement estimation.

Summing up and comparing the diverse methodologies, it can be observed how the execution of large-scale laboratory testing is recommendable. Despite its high cost, it generates a good approximation of the magnitude of the rockfill deformation modulus, which allows one to optimize the design and lower the risks of the system.

With the method of deformation modulus scaling proposed by Saboya (2000), similar values to those obtained by the instrumentation that was installed during the construction were obtained.

The deformation modulus values obtained from the monitoring results and those obtained by means of scaling of laboratory test results are bigger than the ones anticipated for the design by means of literature references; therefore, lower deformations in the concrete face were expected. As a result of this, there was some minor cracking on the first filling of the reservoir with consequent lower seepage flows.

With regards to instrumentation, it is recommended to install redundant types of instruments and amounts that allow cross correlations in order to verify when there is an abnormal behaviour, whether if it's the installation or reading of the equipment and records or if it's presenting a behaviour that requires special attention.

EROSION PRESENTED DOWNSTREAM IN A STRUCTURAL FILL IN THE RIGHT ABUTMENT OF THE URRÁ I HYDROELECTRIC PROJECT (COLOMBIA)

Dam type: Embankment - 73m high
Description: Earthfill with clay core
Case history category: e
Main objective: Early detection of failure mechanisms
Main benefit: Correct and on time remedial action
Observations: Early detection of landslide and assessment of remedial works

A possible slope failure in a disposal fill identified on the right abutment of the Urrá I Hydroelectric Project because of the saturation of materials in the disposal fill. The structural fill was not compromised at all nor the natural slope's integrity on the abutment at this point.

Layout of Urrá Dam

Thanks to the observations from the visual inspections and the proper interpretation of the monitoring results, the risk of failure was identified and mitigated through the installation of a drainage system.

LESSONS LEARNT

For any zone to be used as a disposal site especially if it is adjoining a structural fill, careful consideration should be given to the short and long-term impacts, the compaction methods and placement of the material otherwise future the stabilization costs will increase.

The monitoring system should be properly designed during the design and consider amongst other things soil conditions and parameters to be measured to determine the stability of a structure over a given period of time.

The implementation and execution of preventive maintenance schedules and regular monitoring interpretation are very important for any conclusions pertaining to a structure's safety.

MŠENO DAM – AGEING OF MASONRY DAM FOUNDATION ZONE

Dam type: Masonry dam - 20m high
Case history category: a, d, e
Main objective: Ageing process of dam foundation zone, remedial measures
Main benefit: Bedrock sealing made from new grouting gallery
Observations: Long-term development of seepage regime

Masonry dam was in operation from 1909. A complication during regular maintenance of the area near the downstream toe in the 80´s indicated changes in permeability in the dam foundation zone. There were done several geotechnical surveys to evaluate preciously bedrock, the material in foundation zone and to monitor seepage regime by open piezometers as well. The expert evaluations of the dam stability were made. Operator and dam safety experts carried out large in situ operational tests to verify the influence of hydraulic and temperature loads to the deformation of the dam body.

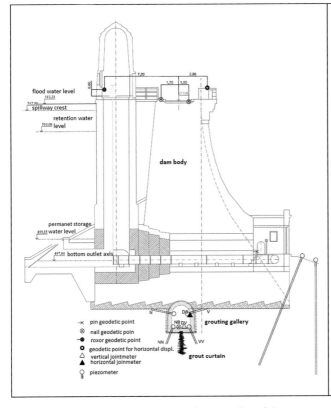

Different ways of remedial measures were taken into account. Finally, more expensive and riskier one was chosen – excavation of the new grouting gallery in contact with the dam foundation zone and short grouting curtain from the gallery to bedrock.

Other advantages of this solution are:

- Detailed geotechnical knowledge of the dam foundation area.

- More reliable monitoring of the foundation zone and grout curtain sealing effectiveness.

- Permanent access to foundation zone for additional sealing works in the future.

Remedial measure works started in October 1998 and finished in December 2000.

Cross section of dam

LESSONS LEARNT

Crucial decision about bedrock sealing technology could be done on the basis of 3 aspects. Very detailed geotechnical and geophysical survey, expert evaluation of dam stability including numerical modelling as well as extra deformation measurements provided during large-scale operational tests. Although the reservoir was emptied to the inactive storage water level during the gallery excavation works, precise dam behaviour monitoring was necessary to control and modify the process and technology of ongoing works.

EL KARM DAM – SPILLWAY AND DAM BODY CRACKS

Dam type: Concrete Gravity Dam – 25 m high
Case history category: e
Main objective: Early detection of cracks mechanisms
Main benefit: Correct and on time remedial action
Observations: Early detection of potential cracks and assessment of remedial works

After seven years of construction of El- Karm dam at the end of year 1998, cracks were observed in the dam body perpendicular to the dam axis and between the dam body (plain concrete) and the stilling basin behind the dam (reinforced concrete). The cracks have been monitored since its appearance It highlighted the utmost importance of dam safety surveillance as well as the significance of proper geological- and geotechnical investigations.

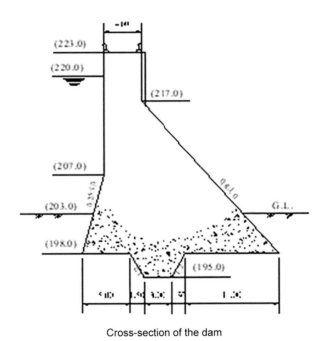

Cross-section of the dam

Thus, the objective of monitoring of the cracks was to stop its movement and blocked it and to prevent the seepage through the cracks.

LESSONS LEARNT

The investigation to know the reason of the cracks produced at the stilling basin walls and floor, finite element has been carried out for the overflow section and concluded that high tensile stress produced at the location of connection between the dam body and the stilling basin wall due to the differential settlement between the dam and the basin. That explains the reasons of the cracks between the dam body and the basin. So the problem was detected and the study and remedial works were followed with the Independent Panel of Experts.

The foundation (bedrock) should be uniform in the longitudinal direction in order to avoid the excessive tensile stress in longitudinal direction and to avoid the differential settlement between the dam monoliths.

The stilling basin of the spillway should be placed separately from the dam body to avoid the formation of cracks between them.

ETANG DAM - UPSTREAM MEMBRANE CRACKING AFTER 20 YEARS OF OPERATION

Dam type: Earth and rockfill dam – 33.5 m high
Case history category: d, e
Main objective: Importance of visual inspection in association with monitoring
Main benefit: Appropriate and early actions engaged for dam safety
Observations: Interest of alarms for early detection of events concerning dam safety

On 22 Feb. 2000, the engineer in charge of dam monitoring was visiting the site, with the reservoir nearly full (1.2 m below Normal WL). He visually noticed that the channel collecting the total leakage was unusually full of water. Immediate readings of the monitoring instruments were done and revealed a significant increase of the total leakage (32 l/s, approximately twice the usual figure), and local increases of piezometric levels (up to + 4 m), behind the upstream membrane and in the foundation.

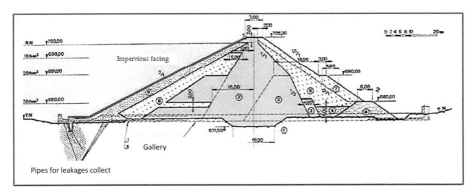

Highest vertical section.

Following the detection of this anomaly, the dam operator decided to draw down the reservoir, at a maximum controlled rate of 0.40 m/hour. During the following visual inspection of the upstream membrane, a crack was found on the upstream toe: it was horizontal, 12 m long, and 5 to 10 mm open. Repair works were immediately undertaken using water reactive resin grouting.

LESSONS LEARNT

The monitoring system fitted well to the physical parameters usually measured for this type of dam: it was designed at an early stage and permitted the analysis of leakages and piezometric levels on some few representative sections.

The anomaly detection was followed by rapid actions and decisions, which may be considered judicious after time: firstly, confirmation of the anomaly with complete hydraulic measurements and first diagnosis, then a rapid dam safety decision (lowering the reservoir water lever), and finally corrective repair works (in two steps).

An essential point is probably the adaptation of the surveillance for this type of dam, in which anomalies may occur very suddenly and rapidly. *Continuous monitoring and alarm systems for the surveillance of hydraulic parameters may contribute in an efficient way to the safety management* in such cases.

Importance of visual inspections and presence of skilled staff for surveillance.

GRAND'MAISON DAM - SLOPE UNSTABILITY ABOVE THE RESERVOIR

Dam type: Earth and rockfill dam – 130 m high
Case history category: a, d, e
Main objective: Monitoring and controlling risks in dam environment
Main benefit: Safe operation of the dam with the slide above the reservoir
Observations: Evolution of measurements (geodesy, photogrammetry, GNSS, InSar)

Grand'Maison dam was built from 1978 to 1985. Since 1981, the stability of the new road along the right bank called for attention, and a levelling monitoring system was installed along 600 m. On May 26th 1986, the reservoir was its final year of impounding, 20 m below the normal water level (Elev. 1695), when cracks appeared in the asphalt of the road. The same day a levelling measurement showed 17 cm road settlement compared to the previous measurement taken in October 1985.

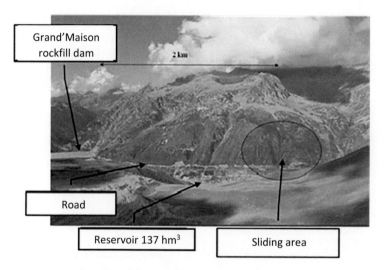

Location of the unstable slope on the right bank

The sliding volume was estimated between 0.5 and 1 million m^3 and could have set off a wave stream resulting in the reservoir overtopping the dam crest. The following measures were immediately taken lowering the reservoir at a rate of 1 m/day, reinforced monitoring, and draining the slope with an 800 m long gallery from which a series of 29 upward drain holes were drilled.

Analysis of the observations attributed the May 1986 slide event as resulting from the increases in piezometric pressures following a very high snow melt rate. Since then, the sliding area is monitored, and settlements have stabilized with maximum values of 5 mm/year.

LESSONS LEARNT

The attention that needs to be brought to identifying and monitoring risks in the dam's environment, right from impounding.

The contribution of monitoring to diagnosis and decision making, both when the events occur and in the permanent management of the risk associated with the slide.

LA PALIERE DIKE - MANAGEMENT OF PHENOMENA DEVELOPING SLOWLY OVER A PERIOD OF 30 YEARS

Dam type: Backfill embankment – 6.5 m high
Case history category: c, d
Main objective: Maintenance linked to monitoring
Main benefit: Keeping dam safety at a controlled level
Observations: Importance of safety assessment for phenomena with slow kinetics and of monitoring after repair works

The structure was impounded in 1982, and in 2005 total consolidation reached 85 cm, which was the maximum settlement foreseen in the design. To maintain the necessary freeboard, a general rising of the dike crest by 50 cm was carried out.

A number of leak zones with a very weak and diffuse flow and without any transport of material (clear water) were recorded during visual inspections at mid height on the downstream slope. At the same time, the piezometers showed a good hydraulic drawdown in the downstream shoulder of the dike. Indeed, these repair works were destabilising events which reactivated the primary consolidation, generated lateral deformations of the structure, and led to a decompression of the dyke body.

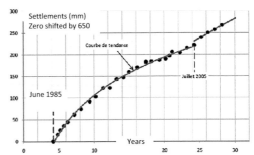

La Palière dike   Consolidation curve at a crest benchmark

After geotechnical investigations, the structure's stability coefficient was verified and found to be satisfactory (F = 1.5), but it will be adversely affected in the event of additional backfilling being carried out.

LESSONS LEARNT

Slowly evolving phenomena are eventually phenomena to which one can become accustomed and pay less attention. In addition to monitoring, the use of modelling and in-depth studies is essential at the right time to better understand the behaviour and the acceptable limits. The difficulty is often knowing when to undertake them. Organisation and people in charge of carrying out and analysing measurements and visual inspections play a crucial role in these situations.

Visual inspections are of primary importance for dikes. It was not the monitoring but the appearance of leaks on the downstream slope which led the owner to undertake a comprehensive diagnosis of the behaviour of the dike and its stability. This study also allowed understanding repair works influence and the importance of modelling their impact before any further operation. Surveillance and maintenance recommendations have been changed accordingly.

MIRGENBACH DAM - SLIDE OF EMBANKMENTS DURING CONSTRUCTION

Dam type: Earth dam – 22 m high
Case history category: a, b, c, e
Main objective: Monitoring water pressures during construction
Main benefit: Controlling water pressure dissipation and embankments stability

In August 1982, when the dam was almost fully built (at 4.30 m below crest level on the upstream side and 3.50 m on the downstream side), a slide was noticed in the central part of the embankment, over a length of about 100 m. The scarp reached 2 to 3 m height after a few days. At the same time, signs of instability were noticeable on the downstream shoulder and on 11 September 1982 a similar rupture occurred there.

Slide of the upstream embankment, 1982.

The measurements recorded during construction showed pore pressures development as the embankment layers were laid. One of the cells (B2) showed abnormally high-pressure levels at the time of the incident. The ratio ru=u/γh between the pore pressure and the weight of the soil was 1.6, considered absurd and faulty. It was attributed to sensor malfunction and held as abnormal.

The investigations showed that unit B2 had not been installed in the planned location. It was at the right level but had not been placed at the location planned by the designer. Thus, the calculation of the ru coefficient took into account a soil weight that was considerably underestimated. The calculation taking into account the real location of unit B2 gave a credible ru value of 0.9, which should have alerted the engineer in charge of surveillance.

LESSONS LEARNT

For large earthfill dams, monitoring is essential during the initial filling and operational phases, but also during the construction phase; this is often left aside.

All measurements showing unexpected or unusual behaviour must be examined carefully and can only be considered as abnormal after a rigorous procedure of analysis or further examinations. In some cases, redundancy of monitoring devices may be useful.

The care taken when installing sensors (in conformity with the design documents) and the control of characteristics and location on the spot after their installation are essential for a valid future use of the measurements.

SYLVENSTEIN DAM - RETROFITTING

Dam type: Embankment dam with clay core – 48 m high
Case history category: d, e
Main objective: Improvement of the sealing system and the seepage flow measurement
Main benefit: Improvement of the dam safety standard
Observations: Innovative methods (deep cut-off wall, additional control gallery)

The dam and subsoil of the Sylvenstein Reservoir were equipped with a new efficient cut-off wall and a reliable seepage water measurement system after operating for more than 50 years. This was the first time in Germany that a 70 m deep diaphragm wall (two-phase plastic concrete cut-off wall), which also cut into rock foundation on either side of the wall, was installed while the dam was still in operation. Monitoring the new sealing system is possible with the drainage piles and an accessible control gallery, which was pressed from the compact rock through the whole dam into the opposite abutment – without hindering the standard operations of the dam, a constructionally brilliant feat that was realized for the first time worldwide.

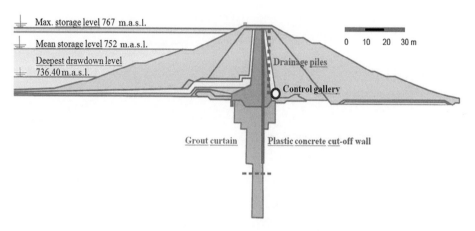

Slide of the upstream embankment, 1982.

LESSONS LEARNT

Bringing a 50-year-old sealing system up to date (clay core / grout curtain in overburden) by adding a cut-off wall to the dam and the foundation (overburden / rock) is possible by using adequate construction methods and machinery (e.g. hydro-cutters).

Improving the dam monitoring and the seepage flow measurement system by adding drainage piles (vertical measure) and a completely new control gallery (horizontal measure) is possible. (Remark: Adding new control galleries should not be considered as a mandatory measure within the scope of embankment dam upgrading projects!)

Under the specific boundary conditions of the Sylvenstein Dam as an important flood mitigation project the dam upgrading measures also have to be considered as preventative measures against the possible consequences of climate change, as the size and small intervals between the recent flood events imply an expected larger toll on the dam.

ALBORZ DAM – OPTIONAL INSTRUMENTATION

Dam type: Rockfill with central impervious core - 78m high
Case history category: a, d
Main objective: Replacement of vibrating wire piezometers instead of standpipes
Main benefit: Longer instrument durability for dam body with high deformation
Observations: Around 60% of standpipes have been distorted due to deformation of the dam body mostly during construction

Around 60% of standpipes have been distorted due to deformation of the dam body mostly during construction. Since vibrating wire piezometers have shown dependable results of pore water pressure over time in many dams, one solution could be reducing the number of standpipes and relying on vibrating wire piezometers to monitor pore water pressure in dam body, especially where the amounts of deformations are estimated to be relatively high.

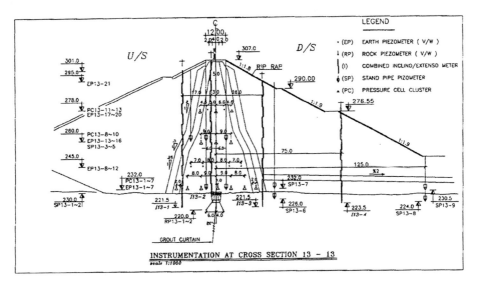

Instrumentation at the maximum cross section of Alborz dam

Above and beyond this, installation details of the standpipes (such as pipe material, sealing grout, casing, etc.) should be considered and done more thoroughly so that they could resist properly against possible deformations of soils without excessive distortions or failure.

LESSONS LEARNT

After a short period of time some of the standpipe piezometers become dysfunctional. Since vibrating wire piezometers have shown reasonable results of pore water pressure over time in many dams, one solution could be to reduce the number of standpipes and to rely on vibrating wire piezometers to monitor pore water pressure in dam body, especially where the amount of deformations is estimated to be relatively high.

Above and beyond this, installation details of the standpipes (such as pipe material, sealing grout, casing, etc.) should be considered more thoroughly so that they could stand firm properly against possible deformations of soils without excessive distortions or failure.

In order to save time and to reduce the costs, one should be more conscientious and vigilant while designing instrumentation systems and consider the number of pressure cells more wisely.

Dam type: Rockfill dam with clay core - 182m high

Case history category: a, b, f

Main objective: Investigation of dam performance during construction, first impounding and the operation period

Main benefit: Adding supplementary instrumentation sections during construction

Observations: Excess pore water pressure in clay core and the maximum amount of pore water pressure coefficient at the largest section of the dam

A continuous increase of piezometric levels in the rock-core contact on right abutment was observed, leading to design remedial works. Grouting program was considered necessary at both the right and left banks and through the complete foundation in order to reduce the potential of clay core piping through open rock joints.

During construction of the dam and before impounding the reservoir, piezometers have shown excess pore water pressure in clay core and the maximum amount of pore water pressure coefficient ($r_u = u/\gamma h$) at the largest section of the dam has reached to 0.42.

It is noted that the dam body piezometers have been affected by reservoir impounding as indicated by the increase of pore pressure in the piezometers of the upstream side of the dam axis.

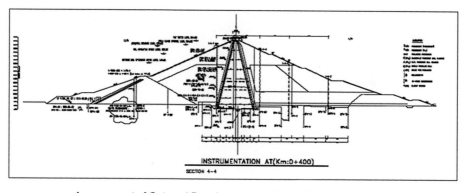

INSTRUMENTATION AT(Km:0+400)
SECTION 4-4

Arrangement of Gotvand Dam Instrumentation on Its Largest Section

During construction, 13 other instrumentation sections were added to install rock piezometers

LESSONS LEARNT

The number of instruments installed at Gotvand dam body and foundation is 1160, including various piezometers, soil pressure cells, magnetic settlement plates, inclinometer and extensometers. By the time of this report to be prepared, less than 3% of the instruments have shown inappropriate results and the remaining have acceptable performance.

Foundation piezometers indicate suitable watertightness of the dam foundation. The clay core piezometers indicate that r_u value is within the allowable range.

The maximum core settlement at the end of construction has been estimated 400 cm at the design stage, while the actual settlement is measured to be 201 cm based on the data obtained from the instruments.

Appraisal of results derived from monitoring instruments indicates that Gotvand dam behaves normal as it was expected.

KARUN 4 ARCH DAM - SURVEILLANCE OF CONTRACTION JOINTS

Dam type: Concrete arch dam - 230m high
Case history category: b, c
Main objective: Early reorganization of dam body cracks
Main benefit: Rehabilitation of some of the cracks
Observations: The trend of cracks opening, and rehabilitation (reinjection) performance (efficiency) has been controlled

During the dam operation (2010–2015) about 132 different cracks have been detected which most of them are capillary. These capillary cracks are formed during dam construction and their openings are not changed after dam impounding. Among those cracks, 6 cracks are important, because of their extended length and opening trend after impounding and also during operation time. Furthermore, the seepage from these cracks is significant.

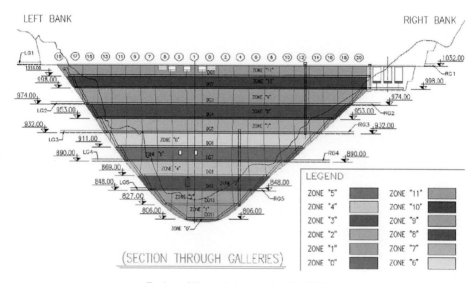

Zoning of Karun 4 dam contraction joints

Injecting of contraction joints of Karun 4 dam body according to the common trend in double arch dams is carried out in two stages.

LESSONS LEARNT

Since impounding of the dam, more than 130 cracks are detected, which most of them are capillary cracks and no seepage are obvious via these cracks.

Just 6 cracks are considered as significant cracks, which have seepage. Their opening and trend of the opening as well seepage during dam operation were considerable. The behavior of these cracks is monitored using crack meters, electrical joint meters and extensometers.

Rehabilitation of these cracks was done based on resin injection in two phases. The efficiency of the injection was controlled by using the mentioned installed instruments based on crack opening tendency and seepage.

The results show the suitable performance of the resin injection for dam rehabilitation. During the resin injection, the operation parameters, such as injection pressure, are modified based on monitoring results.

MASJED SOLEIMAN DAM – BEHAVIOR AND CONDITION OF THE DAM

Dam type: Rockfill dam with clay core - 177m high
Case history category: a, b
Main objective: Controlling dam transient condition
Main benefit: Indicating that the dam behaves consisted but not satisfactory
Observations: Continuous behaviour without abrupt or sudden changes

The interpretations of the measurements carried out in 2006 allow the conclusion that the dam behaves consisted but not satisfactory. The development of measurements shows continuous behaviour without abrupt or sudden changes.

High numbers of damaged instruments are worrying. Some of the damaged instruments have been replaced but these maintenance and substitute works have to be continued.

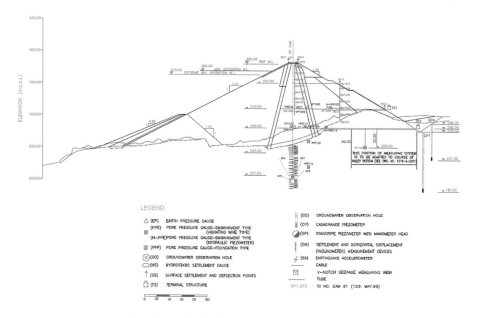

Locations of main instruments Cross section A

After June 2004, Masjed-e-Soleiman dam exhibits large deformations and substantial cracking, but they are acceptable, and the dam core is in reasonable condition. Ongoing monitoring of the key parameters is mandatory.

LESSONS LEARNT

To conclude, the dam seems to be in a transient condition. General it shows consistent behaviour with no abrupt changes but pronounced ongoing deformations and cracking.

The settlements and horizontal displacements are alarmingly high with total values of about 3.6 m surface settlements on the dam crest since construction from 3/1999 to 3/2007 at an ongoing rate of 0.3 m per year and 0.97 m horizontal displacements from 1/2001 to 11/2007 at a rate of 0.09 m per year.

The discontinuation of settlement and inclinometer measurements is negligent behavior endangering the dam safety. It is mandatory to precede settlement and inclinometer measurements of the dam.

Also, the crack openings are alarmingly high and need to be continuously monitored. At the moment, the dam seems in stable conditions, but ongoing deformation and cracking can lead to a sudden substantial degradation of dam safety.

SEYMAREH DAM – INSTRUMENTATION AND MONITORING

Dam type: Double arch concrete dam - 180m high
Case history category: a, b
Main objective: Response of the dam instruments to internal and external factors
Main benefit: Ensuring of the dam instrumentation system
Observations: Main loading of the dam body is thermal

Data analysis of pendulum shows that variations in dam body displacements are mainly affected by the change in ambient temperature during cold and hot seasons of the year as well as the thermal gradient created by the concrete. In other words, the dam body displacements cannot be merely as a result of the reservoir water level variations.

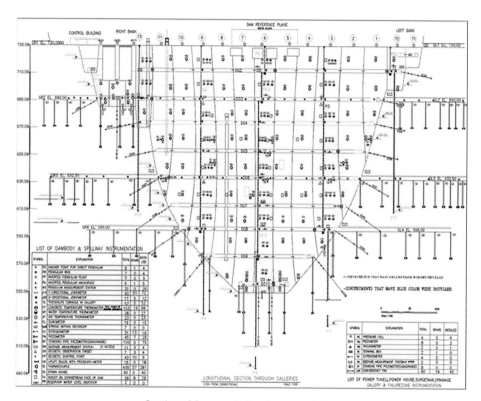

Section of Seymareh dam instrumentation

There is a meaningful behavior between the upstream instruments (installed at the upstream side of the grout curtain), middle ones (between the grout curtain and the drainage curtain) and the downstream ones (downstream from the drainage curtain) and so that the pressure applied at the upstream instruments is expected to be more than that applied at the middle and downstream devices.

LESSONS LEARNT

Considering the maximum elevation ever reached by the reservoir water level (660 m.a.sl.) all of the installed instruments have well responded so far to internal and external factors. Because the reservoir has not yet been completely filled up, the main loading of the dam body is thermal one which has also been confirmed by the data obtained from pendulums and joint meters.

Out of 979 instruments installed so far in Seymareh dam and hydropower plant project, only one electrical piezometer, one thermocouple and one strain meter have pointed out inappropriate readings although a new electrical piezometer has been installed instead. This means that only 0.03% of the installed instruments are out of order which appears to be a good record.

AMBIESTA DAM – CRACKS DETECTION USING SONIC TOMOGRAPHY

Dam type: Arch dam (Dam Height: 58,63 m - Crest Length:144,64 m)
Case history category: e, c, d
Main objective: Detection of cracking patterns inside the dam body
Main benefit: Early detection of cracking patterns variations or extension
Observations:

The investigations have used geophysical techniques such as sonic tomography and micro seismic refraction in the dam wall, with the aim of characterizing the concrete of the area affected by the cracks patterns compared to healthy areas, constituting the dam body, and to trace the penetration depth of the cracks visible from outside face. The sonic investigation has been processed using the modulus of Attenuation of P waves coefficient (MAP).

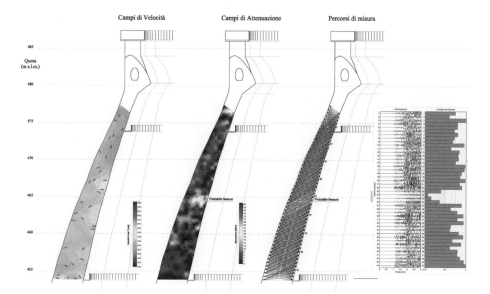

Ambiesta Dam – Velocity tomography (Le) calculus of MAP (middle) and ray path tracing (Right)

The results of the investigation put in evidence that the cracks appeared planar and sub-horizontal towards upstream-downstream, and the cracks are not passing the entire structure. The attenuations of the amplitudes of the individual signal detected have confirmed the presence of the cracks showing a decrease of the energy content of the signals that pass directly through the cracks but highlighting at the same time as the cracks are not through the entire thickness of the dam.

LESSONS LEARNT

The technologies adopted for the investigation put in evidence the ability to characterize the concrete constituting the dam and define the cracks pattern inside the structure.

ISOLA SERAFINI TRANSVERSE - BATHIMETRY TO PREVENT EROSION

Dam type: Transverse
Case history category: e, c, d
Main objective: Early detection of failure mechanisms
Main benefit: Correct and on time remedial action
Observations: Early detection of potential erosion and assessment of remedial works

The monitoring of downstream bathymetry at Isola Serafini transverse is used in order to prevent erosion and verify the rehabilitation intervention, using multi-beam technique.

Isola Serafini Transverse

In order to verify the rehabilitation intervention and to prevent erosion, the dam owner uses multibeam bathymetry technique to define the riverbed downstream of the dam. Such a survey is performed yearly or in any case after a very big flood.

LESSONS LEARNT

The monitoring of downstream bathymetry at Isola Serafini dam, using multibeam technique, put in evidence the possibility to verify the adequacy of the rehabilitation intervention, to control the downstream erosion and to activate subsidiary solution, if necessary.

MISTRAL SOFTWARE FOR ON-LINE MONITORING SYSTEM

Dam type: Different types
Case history category: f
Main objective: Innovative data processing and presentation techniques
Main benefit: to support decision system for dam safety management
Observations:

The use of a software to support decision system for dam safety management based on automatic monitoring system, theoretical reference models and structural behavior. Such installation allows to obtain on-line evaluation, explanation, and interpretation of dam's behaviour, identifying surveillance activities to manage anomalous trends or to minimize critical situations due to flooding or to earthquakes.

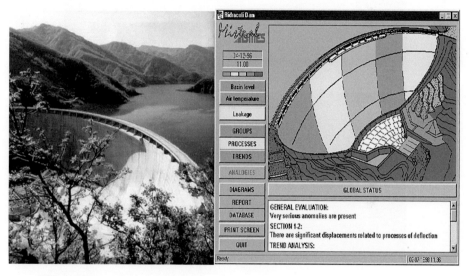

Ridracoli dam and relevant Mistral Interface - general state of the dam (test situation)

The measurements recorded on the dam have been processed to analyze the dam's behaviour. The analysis allowed identifying for each instrument a set of thresholds for the measure and for its rate of variation and confidence limits with respect to the value forecasted by the model. The analysis checked also the logical consistency of the information provided by different instruments affected by the same phenomena.

LESSONS LEARNT

MISTRAL is a decision system for evaluating, explaining, and filtering the information collected by the most important instruments connected to the automatic monitoring system, providing on-line interpretation of the behaviour of the structure in order to support the activity of the personnel responsible of the safety surveillance. Monitoring and data analysis are primary parts in managing the safety of dams by risk assessment methodology. The on-line data analysis and the surveillance management have become a part of the safety procedures of the dams.

SAN GIACOMO DAM – MONITORING OF UPLIFT PRESSURE

Dam type: Buttress (Dam Height: 97,5 m - Crest Length:971 m)
Case history category: e, c, d
Main objective: Stability Analysis
Main benefit: Uplift force to verify sliding condition of the structure
Observations:

The monitoring of uplift pressure at San Giacomo dam allowed to determine the "real" uplift force and consequently verified that the structure is in a safety condition with respect to the sliding condition

San Giacomo Dam

Automatic piezometers were installed to monitor water pressure in the foundation and evaluate the uplift forces at the base of the dam. The piezometer measurements showed that the pore pressure was quite homogenous along the longitudinal profile of the foundation surface. The data collected confirmed the stability of the behavior on time.

The benefits attributed directly to the monitoring program is that the data collected gave the possibility to evaluate the actual global uplift forces acting on the buttresses in comparison to the conventional uplift forces prescribed by Italian Regulation. The measured uplift force confirmed the positive actual safety ratio with respect to stability analysis.

LESSONS LEARNT

The monitoring of uplift pressure at San Giacomo dam allows to determine the "real" uplift force and consequently verify the safety ratio of the structure with respect to static load for sliding condition.

MEASUREMENT SYSTEM FOR AN ASPHALT FACED ROCKFILL DAM (AFRD)

Dam type: Asphalt faced rockfill dam (a pool-type reservoir) – 22.6m high
Case history category: a, d, e
Main objective: Early detection of leakage from the asphalt facing
Main benefit: Easy and quick detection of the position of leakage
Observations: A reduction of repair costs and significant of repair work period

When constructing an asphalt faced rockfill dam in Japan, it is common to divide the asphalt facing into blocks by installing separating layers to prepare for water leakage. Although leakage from the asphalt facing was found during the initial impoundment in the upper reservoir of the Kyogoku Power Plant, the blocking system was highly effective for the early detection of the leaking section, economical repair and the reduction of impact on power generation operation.

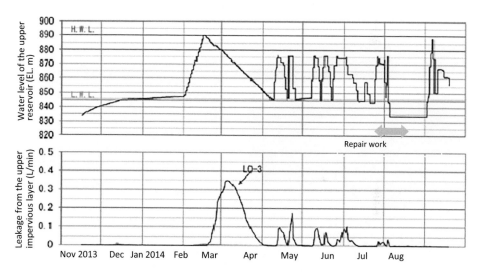

Leakage during the inial impoundment and repair work

LESSONS LEARNT

- The system of dividing the asphalt facing into blocks with separating layers makes it easy to determine the location of leakage in a planar manner.

- Detailed analysis of the changes in leakage with water level fluctuations enables the easy detection of the position of leakage in the vertical direction.

- Easy and quick detection of the position of leakage enables a reduction of repair costs and significant reduction of the repair work period.

AUTOMATIC DIAPLACEMENT MEASUREMENT FOR AN ARCH DAM

Dam type: Arch dam – 61.2m high
Case history category: a
Main objective: Faster delivery of higher-quality information
Main benefit: Significant labour saving and enhanced work safety
Observations: Early measurement of dam displacement after the earthquake

The Okuniikappu Dam is usually unmanned and monitored at a downstream dam using data and monitoring camera images.

The displacement-measuring instruments that allow automatic observation and remote monitoring of measurement data were installed in 1993 with the aims of enabling prompt checking of dam safety in the event of an earthquake, reducing the amount of labor involved and enhancing safety. The theodolites in the instruments were upgraded in 2000 and 2013.

Okuniikappu Dam

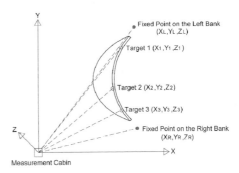

Coordinate System of the Displacement-measuring Instruments

LESSONS LEARNT

When the Great East Japan Earthquake struck Japan on March 11, 2011, the dam site was still covered with snow and was inaccessible by road. However, automated displacement measurement data and camera images helped staff to check that there were no dam abnormalities immediately after the quake.

Before the automation of displacement measurement, it took at least six hours to check the dam's integrity because staff had to access the site on cross-country skis. As avalanches and other hazards tend to arise frequently in March, the site's safety for work was checked using data from unmanned measurement.

DAMAGE TO THE YASHIO DAM CAUSED BY THE EARTHQUAKE AND REINFORCEMENT WORK

Dam type: Asphalt faced rockfill dam (AFRD) – 90.5m high
Case history category: c, e
Main objective: Identification of crack initiation mechanism in asphalt facing
Main benefit: Early confirmation of the stability of the dam body and early reinforcement
Observation: Visual confirmation of damage and reinforcement work from the facing in a short period of time

Immediately after the Great East Japan Earthquake that occurred on March 11, 2011, water leaking from the asphalt facing, which had not been detected previously, rapidly increased to 100 L / min at the Yashio Dam. As a result, the water level quickly lowered and urgent inspections and investigations were carried out. From this, large cracks were confirmed about 20 m from the left and right abutments, as shown in the figure below. As a result of the analysis, it was inferred that the cause of the cracks was the strain concentration on the asphalt facing at the joint openings of the concrete block at the crest. Reinforcement work on the crack sites was completed quickly in time for the summer season when the electricity demand increases, and the water leakage was reduced back to zero.

Yashio Dam

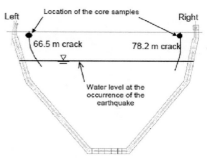

Location of cracks

LESSONS LEARNT

- The leakage from the upper impermeable layer is captured in the intermediate drainage layer, guided to the inspection gallery, and the water leakage detection system makes it possible to identify the occurrence of water leakage and the location of the leakage at an early stage.

- The cause of the cracks is presumed to be strain concentration on the asphalt facing at the joint openings of the concrete block at the crest.

- The residual displacement of the dam body is about 4 mm maximum in both the horizontal and vertical directions, and we evaluate that there is no problem with the stability of the dam.

OUED EL MAKHAZINE DAM – CUTOFF WALL WRONG DESIGN

Dam type: Embankment dam with clay core - 67m high, 530m crest length
Case history category: b, c, d
Main objective: Early detection of failure mechanisms
Main benefit: Correct and on time remedial action
Observations: Early detection of potential erosion and assessment of remedial works

The dam is founded on alluvial deposit where a positive cutoff wall, made of conventional reinforced concrete, has been constructed in the continuity of the clay core. Since the first impoundment of the reservoir occurred in 1979, the pore pressure in the alluvial deposit was gradually increasing downstream of the cutoff wall, while it was decreasing upstream. This was attributed to the continuous deterioration of the cutoff wall as it is brittle and subject to imposed deformations.

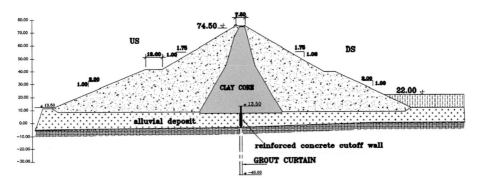

Cross section of dam

The safety of the dam was jeopardized when the piezometric level reached in 1995, almost the downstream toe of the dam with a reservoir 10m below the normal storage elevation. An extensive drainage system including relief wells and deep drainage trenches allowed for the total control of the seepage along with a decrease of the piezometric level at the toe of the dam. Although the seepage flow is still increasing, because of the persistent damage of the cutoff wall, the safety of the dam is not any more threatened.

LESSONS LEARNT

The design for dam safety-monitoring has a robust, reliable and well located instruments in the alluvial foundation downstream and upstream of the cutoff wall. The close follow up and interpretation of the readings based on statistical modelling analysis, permit to detect the increase of pressure and alarm Engineers.

The water level in the relief wells along with the flow collected by the new drainage system are now used to monitor the seepage and pressure in the alluvial deposit.

OUED EL MAKHAZINE DAM – CULVERT JOINT BREAK

Dam type: Embankment dam with clay core - 67m high, 530m crest length
Case history category: b, c, d
Main objective: systematic visual inspection of all works
Main benefit: Correct and on time remedial action
Observations: water stop defects in a culvert crossing a clay core should be considered among the issues to focus on

The dam is provided with a culvert of 15m height and 29m width. It has been used for river diversion during construction. It is divided in two compartments. At the final stage one compartment was used for the spillway and bottom outlet and the other for access and penstock. External walls of the structure are relatively steep. The culvert made of blocs of 10m length each, separated by joints equipped with two lines of waterstops.

In 1984, 6 years after the first impoundment, a dirty leakage appears in a joint of the access compartment. It contains particles of the core. The situation was considered very serious as the reservoir (700 Mm³) couldn't be emptied quickly.

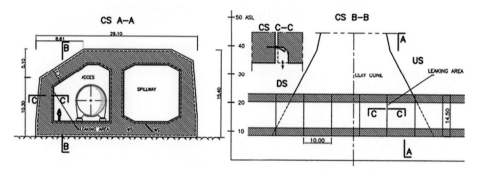

Culvert definition, cross section CC corresponds to the area where the leakage occurs. Almost 30 m3 of core material were washed away

Within 48 hours working around the clock, a thick geotextile has been fixed against the leaking joint, using a 20mm thick steel plate anchored to the structure. Expected filtration was obtained, after the erosion of more than 30 m³ of core material. Few weeks later, it was possible to completely seal the leakage, using sawdust, lentils, and finally conventional grout. Then a systematic grouting of all the joints was conducted to eliminate any risk of new leakage, provided that all culvert joints were stable.

LESSONS LEARNT

Defective joint should be considered as an almost unavoidable hazard during the design stage of a large culvert. Therefore, the number of joints in the footprint of the core should be minimized. In case a joint could not be avoided beneath the core, additional safety measures should be considered beside the double water-stop.

Systematic visual inspection should be extended to all compartment, using adequate lighting and access devices.

Dam type: gravity masonry dam – 15m height on naturel ground and 24m on foundation, 80 Mm³ reservoir capacity. Located in desert low hazard zone

Case history category: c

Main objective: how insufficient design and surveillance are detrimental for dam safety. The owner considering it relatively small dam

Main benefit: strong awareness of the design quality even for dams of limited height

Observations: Geologist expert should be mobilised to check the quality of the foundation, whatever is the type and the height of a dam

The dam foundation is made of hard limestone moderately karstified. A porous, erodible, and continuous marly limestone layer of almost 10m thickness was identified and removed in the footprint of the dam, only in the reverbed where it was outcropping. The geological section along the dam axis is given in the following figure.

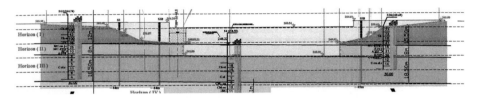

Geological section along the dam axis. Horizon II (pink) is made of erodible marly limestone

During the first impoundment, humidity signs extended humidity signs appeared, in the left abutment and far downstream where the marly-limestone layer is outcropping, evidence of increasing leakage became visible. Few weeks later, uncontrolled flow exceeding 15 m³/s developed as shown in the following photos. Hopefully, the dam did not collapse completely.

A large remedial works was conducted to provide the required safety and water tightness to the dam and its foundation. It includes an extensive grouting and drainage works along with the masonry repair.

LESSONS LEARNT

Whatever is the height of the dam a sound design and a mobilisation of experienced geologist for the inspection of the foundation at the completion of excavations are necessary.

Even in remote areas and low hazard dams, a close surveillance is necessary at least during the first impoundment and operating years.

MURAVATN DAM – PORE PRESSURE CONTROL IN THE FOUNDATION

Dam type: Rockfill dam with central core of moraine till - 77m high
Case history category: d, e
Main objective: Evaluate the nature and extent of problem
Main benefit: Correct remedial action was taken
Observations: Early detection of potential erosion and assessment of remedial works

Muravatn Dam is a 77 m high rockfill dam founded directly on bedrock with a central impervious core of moraine till. It was completed in 1968. An unlined headrace tunnel passes directly underneath the dam as shown in the following figure.

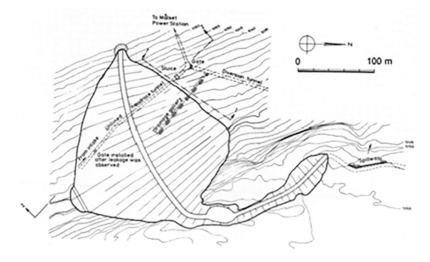

Dam site with location of tunnels

The original monitoring program for the dam and foundation included only one weir station and 68 surface monuments for measurement of leakage and surface displacements. When water was impounded in the reservoir for the first time, a significant leakage of water into an adit shaft downstream of the dam was observed. To properly assess the situation and to plan corrective action, it was necessary to know the water pressure in the fault zone and in the dam foundation. Nine piezometers were installed at different depths in a net of boreholes close to the toe of the dam to monitor water pressure in the foundation. The piezometer measurements showed that the pore pressure was alarmingly high compared to the water level in the reservoir. To control and relieve the high pressure, a drainage gallery was driven into the downstream foundation and a system of drainage holes and observation holes were drilled from the gallery. Pressure sensors were connected to packers installed in eight of the observation holes to continuously monitor pore pressure in order to evaluate the effectiveness of the drainage system.

LESSONS LEARNT

A need for corrective action was confirmed by the measurements of high-water pressure in the dam foundation. Measurements showed the drainage system to be quite effective, and the high pore pressures in the foundation at the toe of the dam dropped radically and to an acceptable value and have remained so ever since. Instruments for the precise and frequent foundation's monitoring during remedial works, is important to follow the remedial process. The Design of the remedial works included an important group of piezometers to check the efficiency of new grouting and drainage curtains and tracing to the future operation and maintenance.

STORVATN DAM – CONTROL OF ASPHALT CONCRETE CORE

Dam type: Rockfill dam with inclined core of asphaltic concrete – 90 m high.
Case history category: c, d
Main objective: Verify new design concept
Main benefit: Design verified
Observations: Construction control of asphalt concrete core

Storvatn Dam is a Norwegian rockfill dam with an inclined core of asphaltic concrete. Construction of the dam was completed in 1987. It has a maximum height of 90 m, a crest length of 1,475 m, and a total volume of about 10 million m³. Most rockfill dams in Norway have moraine cores; however, an asphaltic concrete core was chosen at this site because it was the best alternative. The primary objective of the instrumentation program was to determine the deformations of the dam.

 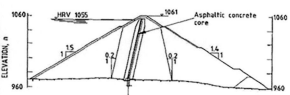

Photograph and cross section of Storvatn Dam

Dense asphaltic concrete is virtually watertight; thus, only a thin membrane is required to prevent leakage. Usually, the width is set at about 0.8 percent of the water head, with a minimum thickness of 40 to 50 cm. In the case of Storvatn Dam the 90 m high membrane varies in thickness from 80 cm at the bottom to 50 cm at the tap. Since it is so narrow, the asphaltic concrete barrier constitutes merely a thin membrane which will follow the displacements of the adjacent fill material. Therefore, the deformations of the asphaltic concrete must be compatible with the movements of the surrounding fill without undergoing any fissuring or crack that could cause leakage.

The main objective of the instrumentation program was to determine the strains that occur in the dam, primarily in the asphaltic concrete core and in the adjacent supporting material. In particular it was desirable to find out if there is any tendency for the width of the core to expand or contract, or to "hang up" on the supporting material as a result of differences in compressibility of the core and surrounding material.

The instrumentation program for Storvatn Dam included 3 instrumented cross sections comprising in all: 3 weir stations; 284 survey monuments; 12 inclined, vertical or horizontal inclinometer casings; 28 extensometers for strain measurements in the asphaltic concrete; 10 extensometers for detecting relative movements between the core and adjacent transition zone; 10 special devices for detecting shear deformations in the transition zone; and 10 pressure cells for measurement of stresses in the rockfill. Special instruments for monitoring strain in the core and shear deformations in the adjacent filter zone were implemented 3.

LESSONS LEARNT

The benefit of this monitoring program lies in the documented deformations of this type of new dam during construction and operation. The data have been used to calibrate the analytical models used in the design of the dam. This information is extremely valuable to the designer of subsequent dams with asphaltic concrete cores.

SVARTEVANN DAM - SETTLEMENTS - DAM PERFORMANCE

Dam type: Embankment dam with moraine core - 129m high
Case history category: c, d, e
Main objective: Appraisal of the stability and performance of a zoned dam
Main benefit: Construction control, long-term monitoring, and validation of the design concept
Observations: Pore pressure build up and dissipation, leakage, relative movements of different materials or zones within the dam

At the time, the Svartevann Dam was built, it differed significantly in concept and size from the existing dams in Norway (40% higher). The design of core was also a compromise dictated by amount of suitable moraine at the site.

The total settlements of the dam were somewhat larger than predicted. Settlements was highest during the first 4 years of operation. Fill material in support zone of Svartevann Dam was not sluiced (washing away the fine material) during placement. The presence of the fines can lead to large initial settlements, especially combined with steep slopes. However, other dams build afterwards with not sluiced support zones, but built with wider slopes, do not show the same behavior. The high initial settlement for Svartevann Dam during the first 4 years of operation can be explained by a rapid building, rapid initial impoundment (40 m) as well as the steep downstream slope and abutments.

Measured pore pressures in the core were modest and dissipated rapidly. The pore pressure reduction through the core of the dam from the upstream side toward the downstream side is concentrated in the downstream side of the core. This trend has also been observed in other dams where a gradient larger than anticipated has been observed in the downstream part of the core. This effect is due primarily to the fact that the core is partially saturated and contains air bubbles that move towards the downstream portion of the core. These air bubbles reduce the permeability of the core which becomes less permeable to flow of water. This anomaly decreases as the air bubbles eventually dissolve in the water flowing through the pores and a stationary flow occurs.

The measured leakage is only approximately 50 % of the original theoretically calculated value, which indicates that the permeability of the moraine in the core material is lower than planned.

LESSONS LEARNT

Monitoring needs to be defined by assessment and the evaluation of constructive safety, surveillance plans and methods, and contingency planning needs. There are no simple rules for determining the appropriate level of instrumentation and monitoring. Monitoring needs will vary depending on the dam behavior considering generic- and potential failure-mode-based performance assessment. Instrumentation and measurements will depend on the type and size of the dam, consequences of a failure, the complexity of the dam, foundation conditions, known problems or concerns and the degree of conservatism in the design criteria. Therefore, selection of appropriate instrumentation and monitoring programs need to be based on the technical consideration along with engineering judgement and common sense.

Technical specifications for monitoring systems and surveillance procedures must provide information on methods of installation and frequency of observations. Action levels for corrective actions and threshold values that indicates a significant deviation from the normal range of readings, must be specified, monitored, and evaluated.

NORWEGIAN TRIAL DAM – VERIFICATION OF NEW CORE DESIGN CONCEPT

Dam type: Rockfill dam with central membrane – 12 m high
Case history category: c, d
Main objective: Verify new design concept for impervious core
Main benefit: Design verified
Observations: Design verification during construction of a new type of impervious membrane

A 12 m high, 120 m long trial rockfill dam was constructed in 1969 with a unique central watertight membrane. This 0.5 m wide barrier was formed in 0.2 m thick layers by filling the voids of a matrix of prepacked rock aggregate with hot bitumen. The cross section and how the membrane was formed in place is shown in the following figure:

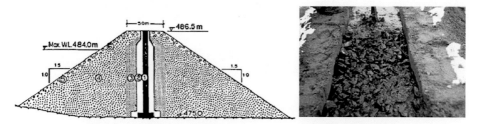

Cross section and photograph of injecting hot bitumen into gravel matrix

In order to prevent the impounded water from "fingering" through the bitumen membrane, the pressure in the bitumen must be higher than the water pressure at a corresponding level in the reservoir. To achieve this, the membrane is extended above maximum pool elevation, thus creating an overpressure in the bitumen relative to the reservoir water pressure.

The primary objective of the measurement program was to measure the distribution of bitumen pressure with depth so that it could be compared with the external water pressure. In addition, it was considered important to monitor horizontal deformations of the membrane since it was constructed of a viscous material.

The instrumentation consisted of survey monuments, six special pressure transducers for measuring the fluid pressure in the bitumen, and extensometers at three levels to measure change in width of the membrane. A piezometer that could be used to measure pore-bitumen pressure had to be developed and tested. Laboratory tests showed that the piezometers worked very well.

LESSONS LEARNT

Measurements were continued on a regular basis for three years after completion of the dam. The measured change in width of the core was very small, of the order of millimeters. The distribution of bitumen pressure with depth was less than hydrostatic with respect to the top of the bitumen, but the pressure measured at the various levels was always slightly higher than the reservoir water pressure at the same depth.

- Satisfactory performance of the new type of membrane was verified.

- The measurements and experience obtained on this dam provided a basis for the design and construction of five other small dams of this type.

VIDDALSVATN DAM - LEAKAGE - PIPING - SINKHOLES

Dam type: Embankment dam with moraine core - 96 m high

Case history category: c, d, e

Main objective: Serious turbid leaks, erosion of the core in the general vicinity of the two sinkholes

Main benefit: Knowledge about the filters

Observations: Early detection of serious turbid leaks

The Viddalsvatn Dam was constructed in 1970–71. Major concentrated leaks of turbid water leaks occurred in autumn and winter 1972. Leaks emerged abruptly several times at the downstream toe of the dam the first time the reservoir was nearly full.

Also, by the second filling of the reservoir, autumn 1973, a concentrated leak was observed and sinking (sinkholes) in the top of the dam were recorded. Before the third filling of the reservoir, the core had been repaired by grouting. In autumn 1980 it was again registered a momentary leakage at 180 l/s.

Photograph of the Viddalsvatn Dam

LESSONS LEARNT

The leakage and erosion have been caused by selective erosion of the fines out of the moraine core material through the coarse downstream filter. Surface manifestation of piping within the core material took the form of an abrupt increase of seepage discharge laden with sediments, and the appearance of sinkholes. The incidents did not involve dam breach but created a real sense of urgency for remedial action and required costly repairs.

The experiences were very similar to a number of other dams of similar core materials, which has been indicating that these materials are internally unstable and require finer downstream filters than previously believed desirable. Conclusion was that there is no possibility of a complete failure of the dam. The worst probable future trouble will be a repetition of the post leaks. It was not necessary to lower the reservoir or take rapid remedial measures when new turbid leaks appear.

A new reassessment was completed in 2014, instrumentation and measurements comply with the regulations. Several other safety deviations have been detected: crest protection, size of placed riprap and freeboard. Measures will be carried out to correct those deviations. Continuous efforts should be made to improve the surveillance systems for dams with imperfections.

ZELAZNY MOST TAILING DAM – DILIGENT MONITORING DATA ANALYSIS

Dam type: Ring-shaped tailings dam with a perimeter of 15 km – 22 m to 60 m high.
Case history category: e
Main objective: Input for the Observational Method
Main benefit: Design modified on basis of measurements
Observations: Observational method for failure mode detection

This case history deals with the Zelazny Most tailings dam in south-west Poland. The ring-shaped dam has a perimeter of 15 km, area of 20 km², and is one of the largest tailings' dams in the world. Approximately 80,000 tons of waste are transported hydraulically to the dam every day. Deposition of tailings started in1975. The current height of the dam is between 22 to 60 m above the original ground surface.

1) Decant pond	4) Slurry pumpout pipeline	7) Monitoring well
2) Beach*	5) Confinement dykes	8) Toe drain
3) Spigot pipe	6) Starter dam	9) Phreatic line

(*) Beach width larger than 200m, usually between 300 and 800m

Aerial view of the tailings dam and cross section of part of the dam

The monitoring program includes measurements of:

- Mining-induced seismicity: 2 seismographs and 10 biaxial accelerometers.

- Pore water pressure and elevation of the phreatic line: 1800 open standpipe piezometers and 300 vibrating-wire piezometers in boreholes. The latter were installed by the fully grouted method (Mikkelsen and Green 2003).

- Surface displacements: 350 benchmarks for geodetic and GPS measurements plus an Automatic Total Station with 23 target mirrors.

- Measurements of subsurface displacements of the dam and foundation: ~50 Inclinometer installations.

The main concern initially was instability due to potential flow liquefaction of the tailings. Therefore, the first inclinometers installed through the dam into the foundation were fairly shallow. However, after a few years of operating the dam, geodetic data and inclinometer measurements showed some sections of the dam were moving more or less as semi-rigid bodies, and deeper inclinometers were installed in 2003. The dam and subsoil above elevation 80 m are sliding along a shear plane in the Pliocene clay at about elevation 80 m. This is at a depth 35 m under the original ground surface. There is also a zone of concentrated shear at about elevation 70 m.

LESSONS LEARNT

The design and operation of the Zelazny Most tailings dam is an excellent example of the use of the Observational method in geotechnical engineering. Measurements have resulted in design changes and remedial measures such as:

- Moving the dam crest upstream to flatten the average downstream slope.

- Constructing stabilizing berms at the tam toe; and

- Installing relief wells in the foundation to reduce pore water pressures.

SAFETY CONTROL OF CONCRETE DAMS AIDED BY AUTOMATED MONITORING SYSTEMS. THE PORTUGUESE EXPERIENCE

Dam type: Concrete dams
Case history category: a, b, c, f
Main objective: Real-time structural safety control
Main benefit: Detect possible malfunctions as early as possible

In Portugal, automated monitoring systems have a conventional structure consisting of automated data acquisition and transmission systems as well as data processing and management modules. A new data processing and management system for monitoring, diagnosis, and safety control, called gestBarragens, has been developed and used by LNEC since the early 2000's, in accordance with LNEC's assignments as technical advisor of the Dam Safety Authority. LNEC has performed and promoted scientific and technological research works and the development of the gestBarragens system, aiming to the continuous improvement of the activities of structural safety control, summarized in three questions:

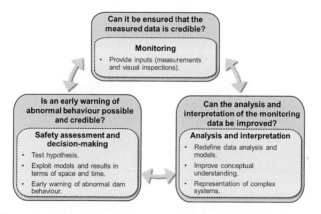

Main questions address about safety control of concrete dams aided
by automated monitoring systems.

New methodologies were developed in order to: i) improve the reliability of the measured data provided by automated data acquisition system, through the implementation of a quality control routine for the validation of the data measured, taking into account the measurement uncertainty of the systems; ii) improve the routines related to data analysis and its interpretation by proposing the use of data-based methodologies in order: to support the decision of which physical quantities should be automated, and to allow the pattern recognition of the structural dam behaviour; iii) define decision rules for early warnings related to the identification of abnormal behaviour.

LESSONS LEARNT

The safety control of concrete dams, based on monitored data, continues to be an actual and important challenge being faced by structural engineers. The main concern, for detecting possible malfunctions as early as possible, is the real-time assessment of the structural behaviour under operating conditions. For an effective real-time decision, confidence in the measured data is crucial. Additionally, it must be possible to interpret these data (through adequate data-based methods) in order to properly assess the structural behaviour and condition (with the support of reliable numerical models). Consequently, it is fundamental to provide to the entities responsible for dam safety a management information system to allow data access, interpretation of the information and decision making, as quickly as possible.

GURA RÂULUI DAM – LONG TERM BEHAVIOUR ANALYSIS USING STATISTICAL AND DETERMINISTIC MODELS

Dam type: mushroom-shaped buttresses, 73.5 m high

Case history category: e, c

Main objective: Early detection of potential failure of the foundation

Main benefit: Constructive interventions only if necessary

Mathematical models used: statistical models of EDF and deterministic finite element models for behaviour assessment

The bedrock is made of amphibolite gneiss. The geological structure is characterized by the high slope of the schist deposit (50–80 degrees) and by multiple fractures and cracks.

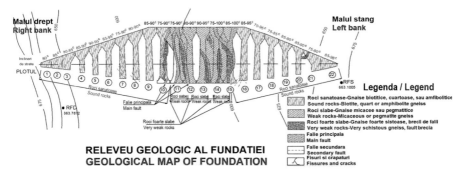

RELEVEU GEOLOGIC AL FUNDATIEI
GEOLOGICAL MAP OF FOUNDATION

The Gura Râului Dam – Geology of the site

The assessment of the dam behaviour was performed by various methods of modern analysis: statistical models of EDF and deterministic finite element models. The behaviour models developed at the Gura Râului dam provide the opportunity to get the structural response for different combinations and stress levels, being used for interpreting the collected data, the regular evaluation of dam safety and predicting future behaviour. "Dynamic limits" of statistical models of the displacements measured at pendulums and rockmeters had been implemented in the UCCWAT application. The advantage of using "dynamic limits" versus threshold-type limits is obvious, because these limits check the seasonal evolution of the monitored phenomenon and the normality range is both narrower and dependent on the probability of complying with the statistical model.

The deterministic model developed to simulate the response of the dam-foundation system to environmental action is a numerical model with 3D finite elements. The data from the free vibration measurements were used for the calibration of the model. The periodical determination of the dynamic characteristics of vibration of the dam aims to detect the unchanged state or degradation, as an aging of the structural integrity, depending on the evolution of its own periods.

LESSONS LEARNT

The early identification of the foundation soil deficiencies in conjunction with effective constructive measures and the further follow-up of the proper behaviour of the dam is the guarantee of a safe operation of the asset. In this case, the owner focused on monitoring the behaviour of the dam by promoting several modern analysis methods of the data, derived from measurements taken with the instrumentation: statistical models of behaviour type EDF and deterministic finite element models which were calibrated with vibration measurements.

PALTINU ARCH DAM. INCIDENT DURING FIRST IMPOUNDMENT

Dam type: Arch dam
Case history category: e
Main objective: early detection of failure mechanisms
Main benefit: Correct and on time remedial action
Observations: Early detection of failure mechanisms and evaluation of remedial works

During first impounding, in June 1974 while the water level reached a maximum of 642.85 m ASL an increase of the level in the drain drillings on the left slope was recorded, and subsequently, the start of water discharging at some drillings, initially with suspensions and foundation rock displacements towards downstream. These developments were located especially on the left slope terrace but present also in the area of the central blocks of the dam:

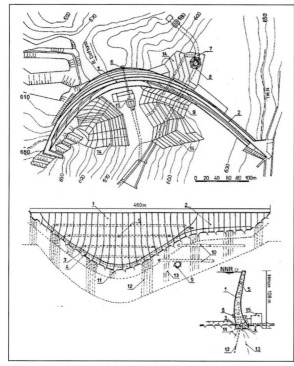

Due to the extent of the abnormal phenomena and the impossibility to control the level by the existing bottom outlet, the storage had to be completely emptied by exploding the plug of the diversion gallery and substantial rehabilitation works had to be carried out.

They mainly consisted in extension of the discharge capacity, improvement of the seepage control at both slopes, drainage works of the space between the grouting curtain and the existing F2 fault, decrease of the bank deformability by execution of a concrete massive on the left slope and a concrete protection on the right slope, supplementing of the surveillance system. The reservoir reached again the Normal Operation Level in 1995.

Lay-out, downstream view and cross section of Paltinu Dam

LESSONS LEARNT

An accident or even an incident is always preceded by atypical phenomena that seem not having any implication upon the dam safety. The quick and correct analysis can have important consequences in the subsequent operation.

A succession of less impervious zones can lead to the increase of pressure in the space between them and, without an adequate drainage to an accident.

Learning the causes of an incident can be carried out by numerical modelling and comparing the results with the measurements performed during the incident. The correct assessments of the causes can lead to an effective and economic remedial solution.

Mastering the situation at the moment when an incident occurs depends on the possibility to control the reservoir water level, therefore on the discharge capacity of the facility.

PECINEAGU DAM

Dam type: Rockfill dam with concrete face – 105 m height
Case history category: b, f
Main objective: a rational and safe use of reservoir for water supply and power generation
Main benefit: Correct and on time remedial action
Observations: Early detection of potential erosion and assessment of remedial works

The Pecineagu dam is a rockfill dam with concrete face, and it has a slope of 1:1,717 and, in the horizontal plan, it has a curvature with a radius of 2000 meters at the crown. The sealing mask has a variable thickness (1.20 ÷ 0.30 m), being made of concrete. It consists of 297 tiles with plan dimensions between 7x10 and 15x10 m². It rests on a massive concrete cut-off wall containing a perimetral gallery for injections and for monitoring.

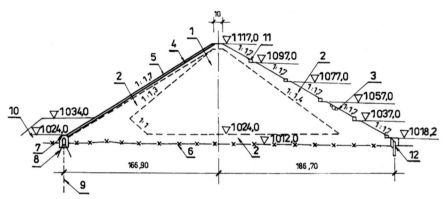

Type-profile of dam:

1 - rockfill type I.	7 - upstream toe wall.
2 - rockfill type II.	8 - Consolidation grouting.
3 - manualy arranged rockfill.	9 - Grout curtain.
4 - concrete face support layer.	10 - Clay prism.
5 - water sealing face.	11 - Instrumentation cabins.
6 - foundation rock.	12 - Downstream cut– off wall.

The dam was commissioned in1985 and the filling has been carried out gradually, having reached maximum levels that increased by about 10 m from one year to another, but as the reservoir level rose over 1080 m, an increase of the infiltrations was recorded, much more than the previously recorded values, at similar levels.

The inspection of the concrete facing, (visual and with georadar) emphasise thewhen the outstanding faults of the perimetral tiles that are resting on the bedrock.

Repairment works consist in a geo- membrane applied on the concrete face.

LESSONS LEARNT

Deformations measured on the mask explain the degrading mechanism of the perimetral tiles and draw attention upon the fact that they can be very large also at the lower side of the dam which is not visible.

Relation between the infiltration flow and the water level in the reservoir draws attention on overlaying of a diffuse infiltration and a multitude of spread local faults and it justifies a radical action.

Upstream vie aer applying the geomembrane on the concrete mask

POIANA UZULUI BUTTRESS DAM

Dam type: Round headed buttress dam - 80 m height - went into service in 1970
Case history category: c, e
Main objective: a rational and safe use of reservoir for water supply and power generation
Main benefit: safety in operation
Observations: Early detection of potential displacements and increased drainage phenomenon

The buttresses present a few constructive particularities: the rock foundation is made up of alternating sandstones and schists; the buttresses are enlarged toward their lower end, creating in this way a continuous foundation; the space in between the buttresses is filled, improving the stability against sliding; „T" shaped upstream, invert-less, galleries have been constructed along the plot joints; drains are extending from these galleries.

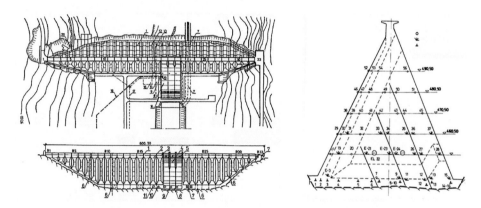

Poiana Uzului Dam. Lay-out plan, downstream face and typical plot.

At the end of April 1984, at the same time with the water level rising up to 510.48 m ASL which anyway stays below the maximum reached before of 513.40 m ASL, the behavior abnormalities amplify: abnormal movements of the right bank plots (plots numbers 5 to 10); very high flow rates of the drained water and sudden functioning of draining boreholes which were inactive in the past; direct infiltrations at the contact line between the concrete and the rock in the drainage galleries in between the plots. Because of the plots movements in the downstream direction, cracks appear on the dam crest and the joint between plots 6 and 7 is being opened.

Remedial works: 205 drainage boreholes, additions to the monitoring system (100 boreholes for piezometers and 12 rock meters) and grouting injections of the bedrock at the bottom of the right shoulder plots (7 rows of 20 meter deep bore holes, at the bottom of the plots 6 to 12), a new method to analyze data of monitoring system.

LESSONS LEARNT

An accident or an incident is almost always preceded by atypical phenomena which seem not to have effects on the safety of the construction works. A quick and correct analysis may have important consequences on the later operation.

At buttress dams, the influence of the water pressure and of the variations of the temperatures on the strains and deformations has the same order of magnitude.

For the bedrocks made up of successive sandstone and schist layers, the cement injection does not create every time a continuous, impervious grouting curtain.

The analysis of the information provided by the monitoring system helped in establishing an empirical relation for quantifying the effect of overlaying water level and temperature in creating upstream stresses that can open seepage paths.

BELFORT DAM REHABILITATION: WALKING THE TIGHTROPE

Dam type: 17m high composite dam: buttressed arches flanked by earth-fill embankments
Case history category: e
Main objective: Early detection of failure mechanisms, control of remedial works
Main benefit: Correct remedial action
Observations: Detection of cracks and ground deformation

During 5-yearly dam safety evaluations of Belfort dam the crack pattern on the downstream face indicated that the principal stresses are in opposite directions than expected for an arch. Tensile cracks up to 20 mm wide were discovered on the upstream face. This was ascribed to a typical AAR crack pattern until significant foundation movements were discovered when the foundation of the apron downstream of the right arch was cleared during the rehabilitation works. The obstacles discovered during the design and construction of the rehabilitation works and how these challenges were addressed, and lessons learnt are briefly described.

Crack in arch spillway and relative movements of the rock in the foundation of the apron next to the right abutment

LESSONS LEARNT

Not to jump to conclusions. The crack pattern of the arches was initially interpreted as severe Alkali Aggregate Reaction. The possibility of horizontal translations around the toe of the abutment walls were not considered based on the fact that the side walls were plumb and straight in line when sighted in an upstream-downstream direction. That was a wrong conclusion when foundation movements were detected.

Less can be more. The lean design and construction teams proved to be very effective especially as far as quality control is concerned, both parties took responsibility. Site meetings were short and to the point. Site supervision staff (the design engineer and a technician) spent most of their time on site and not in offices. Site supervision staff performed had performed all relevant construction tasks in order to be in a position to evaluate and supervise construction activities

Importance of good teamwork and team spirit within each team and between the teams. The construction team was involved in the design right from the onset. Construction valued the fast reaction time from the design team

Be careful with grout pressures. Hydraulic fracturing may occur faster than one thinks.

DRIEKOPPIES DAM MONITORING

Dam type: Embankments flanking gravity spillway - 50 m high
Description: Zoned earthfill with clay core and concrete
Case history category: a and e
Main objective: Early detection of failure mechanisms
Main benefit: Correct remedial action, control of construction and remedial works
Observations: Hydraulic and pneumatic fracture, drilling in embankment, pressure cells

The design and installation of the instrumentation system of Driekoppies Dam established a good practice for drilling into embankment dams and their foundations as well as the configuration of pressure cells. Pneumatic and hydraulic fracturing were also monitored during construction.

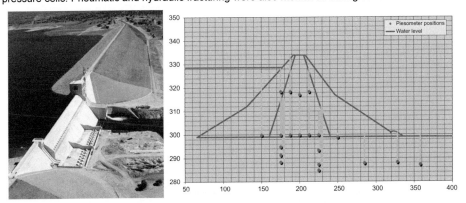

Dam aerial view and pore pressure distribu on at chainage 7410 in 2002

LESSONS LEARNT

Importance of a well-designed and well-installed monitoring system. The system should be designed with sufficient redundancy. Sufficient time should be allowed for installation.

Drilling into embankment or its foundation for installation purposes should be done under supervision of the instrumentation team. For pneumatic drilling, the air pressure at the tip of the drill should be regulated in order to blow the material out of the hole without fracturing the soil.

Earth pressure cells can be very useful if they are properly designed and installed in clusters. Pressure cells installed in in a single cell arrangement cannot be interpreted due to lack of all relevant stresses, i.e.in all three dimensions.

Single earth pressure cell installations without piezometers are problematic because they do not give the full picture.

Jumping to conclusions should be avoided when interpreting and evaluating instrumentation results. The pneumatic drill operators were falsely accused.

Monitoring is an interdependent chain of activities: Independently on design and installation, monitoring systems have to be maintained and the results evaluated regularly.

The relatively short lifespan of digital equipment: While electronic components may still be available, incompatibility with computer equipment repeats itself every 10 years or so.

INYAKA DAM MONITORING

Dam type: Embankments flanking trough spillway - 56 m high
Description: zoned earthfill with clay core and concrete
Case history category: a, b, c, e and f
Main objective: Control of contact between embankment and concrete structure
Main benefit: Obtaining pressure distribution and main stresses
Observations: Design and installation of pressure cells, graphical representation of data

The monitoring of Inyaka dam benefitted dam monitoring and dam safety surveillance in more ways than one. It highlighted the correct layout, installation, and evaluation of total pressure cells as well as useful techniques to display the results of pore pressure gauges and total pressure cells.

Dam aerial view and layout of earth pressure cells

LESSONS LEARNT

It is essential to evaluate total pressure cells by means of principal stress vectors. (The use of commercially available software packages in this respect cannot be over emphasized).

As far as the use of earth pressure cells are concerned sufficient redundancy must be provided.

Recorded "failures of total pressure cells" may very well be as a result of lack of redundancy or failure in the understanding of what is being measured.

Failure of many instruments is actually installation failures. One of the authors made the following general comment at ICOLD 2000 during the discussion on total pressure cells: "... instruments are more and more installed by less and less experienced technicians following outdated practices perpetuated in literature and supervised by completely inexperienced engineers. Therefore, I think most instrument failures are really installation failures".

Designers of monitoring systems should as far as possible be involved in all the processes of monitoring, from design up to evaluation. This was the case at Inyaka dam.

Not only the installation of instruments but the effective presentation of the results is equally important parts of the monitoring system.

The concept of appropriate dam safety-monitoring boils down to using reliable and well-located instruments that is carefully installed, diligently observed, and properly evaluated.

KATSE DAM MONITORING

Dam type: Arch - 185 m high
Description: Double curvature concrete
Case history category: a, b, c and e
Main objective: Early detection of failure mechanisms
Main benefit: Detection of deformations and cracks
Observations: System performance of monitoring system due to installation

The monitoring system of the highest dam in Southern Africa viz., Katse Dam in Lesotho, was well designed, but the installation of the monitoring equipment (to the contrary) had to be carried out under adverse conditions. A 1 000 m³ slope failure (years after completion) on the right flank as well as the results of the TRIVEC installation, pendulums and extensometers in the immediate vicinity are briefly discussed.

Downstream face of dam and horizontal displacements measured with TRIVEC

LESSONS LEARNT

The adverse effect of a not so well-installed monitoring system has been underlined. The installation of the monitoring system at Katse Dam is not perfect, but at least a good example of how it should not be done.

Ample redundancies in the design of the monitoring system, counteracted the sub-standard installation.

The TRIVEC emerged as one of the best scientific monitoring devices for 'forensic engineering studies. The phenomenal quality of the results clearly demonstrated its value to determine the behaviour of the dam's foundations and to predict potential problems.

Cost of both TRIVEC observations and Geodetic Surveys are relatively expensive. They are therefore done at 6-monthly intervals or whenever the need arises. At Katse dam the measurements and evaluation report for the TRIVEC system (12 holes and 800 measuring points) requires 100–200 man-hours and the Geodetic Surveys 300–600 man-hours for a full measurement.

Pendulums are excellent devices for day-to-day dam safety monitoring. Geodetic Surveys provide a larger picture of the deformations at the dam wall relative to a network of beacons around the dam. The TRIVEC on the other hand enhances the readings of the pendulums by providing detailed (3-D) information along the length of the borehole. The three different systems complement each other.

KOUGA DAM MONITORING: SOLVING THE RIDDLES

Dam type: Arch - 72 m high
Description: Double curvature concrete
Case history category: a, c and e
Main objective: Detection of swelling and structural deterioration
Main benefit: Understanding the influence of dynamic behaviour on structural deterioration
Observations: Analysis with dynamic AVM monitoring and detection of AAR

The instrumentation system of Kouga Dam was regularly updated to answer specific questions. Swelling action in the concrete detected soon after completion was only 10 years later confirmed as AAR. Several instruments were added, as their need arises, such as 3D crack gauges, Sliding Micrometers, TRIVEC, improved Geodetic Surveys and finally continuous monitoring using GNNS/GPS for static and AVM for dynamic monitoring.

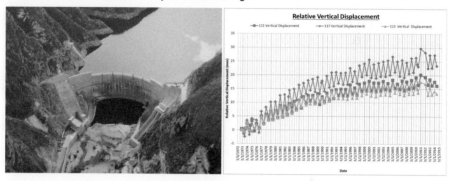

Dam aerial view and vertical displacement during 40 years of operation

LESSONS LEARNT

The importance of AVM monitoring over long periods of time in order to identify possible structural deterioration has been demonstrated. The one AVM done during low water levels assisted to get a better picture of the behavior of the dam.

Value of tri-axial dynamic measurements versus uni-axial or bi-axial measurements has been clearly proven by the AVM monitoring at Kouga dam.

AVM measure micro movements while with strong motion measurements relatively larger movements are measured and scale effects must be taken into account.

Fluid interaction with the structure seems to play a minor role during AVM in contradiction to what was previously suspected. This is not the case with strong motion movements where the fluid interaction does play a major role.

Continuous AVM monitoring provides valuable information for scientific purposes especially during strong motion vibrations. A roving system may seem to be (cost-) effective.

Ambient vibration testing has the advantage that dynamic properties are measured under the structure's operating conditions. Dynamic testing using AVM as source of vibration can offer vital links between the actual behavior of the structure and the finite element models.

Measuring translations and rotations across cracks or joints, such as 3D Crack-Tilt gauges, provide a better picture of joint behavior than only translation measurement.

OHRIGSTAD DAM GEOPHYSICAL INVESTIGATIONS WITH A TWIST

Dam type: Embankment - 52 m high
Description: Concrete faced rockfill
Case history category: c
Main objective: Leak detection
Main benefit: Detection of flow paths
Observations: Results from bio-location and other techniques matched

The value of Bio location was discovered during extensive geo-physical investigations to determine the flow path of leakages through the CFRD Ohrigstad Dam. Different methods like visual inspections with divers, hydrophones, fluorescent tracers, self-potential method and bio-location was used.

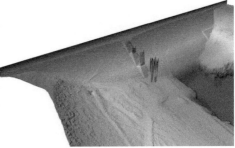

Dam aerial view and underwater survey of Ohrigstad Dam

LESSONS LEARNT

The term bio-location is presently preferred to describe the method. Several terms have been used since 1982, amongst others, water divining, geo divining, geo location etc.

Value of bio-location as a fast "geo-physical" investigation technique. The results of a few minutes geo-divining survey were intended to be used as a humorous ending for the presentation of months of geophysical test results in 1979 to determine the seepage paths through the dam and its foundation. When time came to use these water divining results in 1982, it was discovered the water divining results in fact tie in well with the results of the other methods. What was supposed to be a humorous ending to the presentation backfired. The method became a valuable tool.

Bio-location is not a recognised scientific method, despite the fact that it is acknowledged and used by many "scientists" (engineers, geologists etc.). The method is therefore not for the full-blooded nor the feint-hearted scientist. However, the former Soviet Union did a lot of their ore exploration by means of bio-location) have used it successfully in dam safety evaluations for several decades.

LA ACEÑA DAM - DGPS REAL TIME MONITORING

Dam type: arch-gravity dam – 65 m high
Case history category: a, c, d, e
Main objective: Validation of the DGPS system
Main benefit: Useful in dam monitoring and safety programs, valuable complement to other monitoring methods and interesting for dams with difficult access
Observations: Detects absolute deformations and millimeter accuracy. Affordable cost

With the development of new technologies, computing and telecommunications, new displacements auscultation systems have emerged. The motion control system based on DGPS technology applies a statistical filter that sensitivities are achieved in the system to ± 1 mm, sufficient for normal auscultation of dams as required by current regulations. This accuracy is adapted to the radial displacement of dams, which are common values in coronation amplitudes up to 15 mm in gravity dams and up to 45 mm in arch or arc dams.

This research aims to analyze the feasibility of DGPS system in controlling movements of concrete dams, comparing the different systems auscultation.

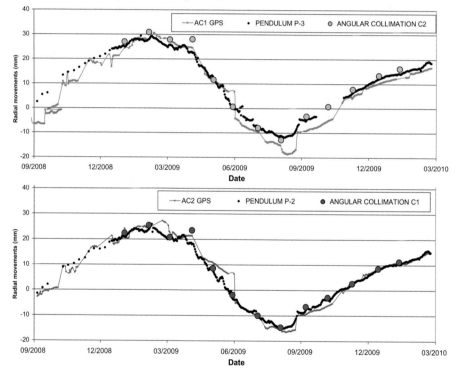

Results comparison: DGPS, pendulums and angular collimation

LESSONS LEARNT

The annual amplitude of observed movements (40.86 mm) is measured well by both DGPS and precise angular collimation. The observed error is between 1,0 and 1,5 mm.

We can affirm that the tested DGPS system is very useful in dam monitoring and safety programs, as it accurately detects absolute deformations and serves as a valuable complement to other monitoring methods.

CASPE II DAM. DETECTING ANOMALOUS BEHAVIOR OF PORE PRESSURES IN EARTHFILL

Dam type: Zoned embankment dam with clay core – 56 m high
Case history category: e, c.
Main objective: Early detection of failure mechanisms
Main benefit: Risk reduction measures and on-time remedial action
Observations: Early detection of potential erosion and assessment of remedial works

Since construction finished in October 1987, several incidents have been detected due to seepage through the river valley's substrate, which had to be corrected by numerous campaigns of remedial works, consisting mainly in waterproofing grouting works as well as a continuous upgrading of the monitoring system, mainly with piezometers (vibrating wire and open standpipe) and seepage control.

At the end of March 1990 during impounding, the seepage control weirs detected an exponential increase of flow at left abutment. The seepage increase was earlier related with a gypsum dissolution process, but it was later proved to be related to foundation and dam internal erosion processes.

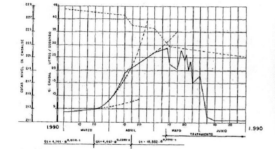 

Evolution graph of seepage and reservoir levels in 1990 and sinkhole in crest.

LESSONS LEARNT

Piezometers and seepage control have anticipated several dam safety incidents at Caspe II dam throughout the years, always related to internal erosion processes in the foundation and dam body. The dam has undergone major repair works and these have affected the performance of piezometers, hence it has been necessary to upgrade the monitoring system alongside.

This dam requires close monitoring system with frequent readings and diligent data analysis in order to detect any behavior change and prevent a progressive internal erosion process on time.

CORTES II DAM - MONITORING AND STABILIZATION OF RESERVOIR SLOPES

Dam type: arch gravity dam – 116 m high
Case history category: e
Main objective: Reservoir slide monitoring in real time
Main benefit: Early warnings in case of detection of movements
Observations: Monitoring of slopes around reservoirs is essential for detecting problems

The quarry for extracting aggregates for the dam construction was initially excavated in the lower part of the old landslide. Cracks were noticed in some silos and retaining walls constructed on the colluvial top part of the slope.

During 1986 and 1987 several actions were taken to find out more about the geology of the affected area and the real extension of the movement as well as its velocity and direction.

This data allowed establishing a procedure of stabilization consisting in removing material from the excavation in the neutral zone to get stable slopes. The material from the excavation was placed in the lower part of the slope. About 800.000 m³ of material was removed from the top part of the sliding slope and replaced in the lower part.

 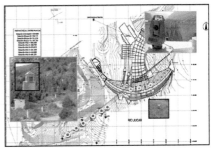

In 2003, the dam owner Iberdrola implemented an automatic control system of the quarry area, consisting of a robotic station that monitors the movements. The system is controlled remotely by the Dam Surveying Service with headquarters located in Valencia

LESSONS LEARNT

Monitoring of slopes around reservoirs is essential in detecting problems and defining solutions.

The responsible dam engineers were seriously concerned about a potential landslide during the first impoundment of Cortes dam. Monitoring of reservoir slopes by inclinometers is an economic and reliable method providing valuable information, both for understanding the sliding mechanisms and for designing solutions for mitigation

Monitoring systems are continuously improving and automation today allows systems to automatically execute remote measurements that help us to monitor the structures and to be alerted at an early stage to future problems that may occur.

ECONOMIC BENEFITS DUE TO AUTOMATION OF MONITORING SYSTEMS LA MINILLA AND GERGAL DAMS

Dam type: Gravity dam 70 m high / Arch gravity dam 63 m high
Case history category: d
Main objective: Upgrading of monitoring system
Main benefit: Cost and time reduction
Observations: Partial automation of readings of monitoring devices

This case history describes the upgrading of the monitoring system of two dams: La Minilla and Gergal, located in the province of Seville (Spain) and belonging to the Metropolitan Water Company of the city of Seville (EMASESA). The whole system also includes the control of hill slopes and the structural monitoring of water storage infrastructures.

La Minilla Dam was commissioned in 1956. It is a 70 m high concrete gravity dam with a storage capacity of 57.8 million m^3. All readings were manual, except the piezometers. Gergal Dam is a concrete gravity arch dam, commissioned in 1974. Its height is 63 m above foundation and the reservoir volume is 36 million m^3. All readings were done manually. Further manual measurements were taken to monitor slopes and adjacent installations.

The management of this system was carried out by manual procedures with handwritten paper form sheets for the readings, typed into Excel spread sheet files for data analysis and storage. The number of man-hours devoted to this purpose and therefore the costs were very high. In order to improve the quality of the process, to reduce costs and to make the system more flexible it was decided to investigate possible solutions comprising automatic data acquisition, supported with the most adequate computer application to manage the volume of information to be handled.

There is a wide range between the existing manual monitoring network and a completely automated system, in terms of technology as well as regarding the costs. Therefore, three solutions where compared: 1) a Fully Automatic Solution, 2) a Partially Automatic Solution and 3) an Advanced Partially Automatic Solution. After a technical and economic study, the Advanced Partially Automatic Solution was selected. Additionally, to its economic advantages the adopted solution maintains a closer contact with the reality of the monitored works and therefore permits a better control of them.

The solution has recently been implemented in its first phase and is currently successfully fulfilling planned expectations.

LESSONS LEARNT

Existing dams, especially the older ones, are usually exclusively equipped with monitoring devices that require important human resources to be read and maintained. These monitoring systems have limitations in terms of increasing the data reading frequency in emergency situations, may have a limited reliability, and data storage and processing is slow and difficult.

In such a case, there might be a need to innovate and upgrade the monitoring system of the dam. The problem that arises is the selection and design of a system that balances system efficiency with economy of the employed resources. Even if the economic possibilities and also the level of human and material resources are high, masses of information do not necessarily provide a better understanding of the behaviour of the dam.

DETECTING ANOMALOUS BEHAVIOR OF SEEPAGE IN EARTHFILL DAMS: LA LOTETA DAM (SPAIN)

Dam type: Zoned embankment dam with clay core and gravel shoulders, upstream clay blanket and cut-off wall – 34 m high

Case history category: e, c

Main objective: Early detection of failure mechanisms during first filling

Main benefit: Detection of a failure mode, follow-up, and implementation of on-time remedial action

Observations: Early detection of seepage avoiding potential erosion and construction of an impervious cut-off at left abutment

The dam was completed in 2008 and during first filling, seepage increased exponentially; reaching 45 l/s, and soil particles were detected in the drainage system, at the left abutment. Dissolution and karstification of the gypsum layer within the foundation are the major risk factors. Control of pore pressures, piezometric levels and seepage (plus dissolved solids) will always be a must at all stages of the dam.

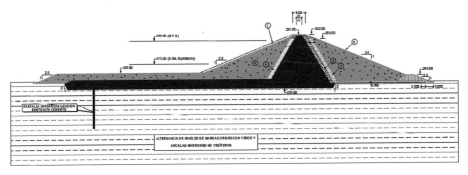

Cross section of La Loteta Dam at profile P-10.

During 2015–16, an additional concrete cut-off wall was built next to the left abutment, connecting with the original cut-off wall. This new cut-off wall is 100 m long, 0,65 m wide, and it is 31 m deep. Currently, the reservoir has resumed operation and it is being filled up again.

LESSONS LEARNT

Monitoring of pore pressures, piezometric levels and seepage (plus dissolved solids) in the foundation, dam body and abutments are of paramount importance during the construction, first filling and operation of the reservoir.

Visual inspections, hydraulic monitoring, a thorough analysis of monitoring records by qualified engineers, plus having a reliable data recording system are key elements in the dam safety management of La Loteta Dam.

SIURANA DAM – OPTIMAL METHOD FOR REMEDIAL WORKS ON DRAINAGE SYSTEM

Dam type: Concrete gravity dam - 63m high
Case history category: c, d, e
Main objective: Improving dam safety allocating minimum investment
Main benefit: Remedial and efficient works with high-cost reduction
Observations: Control and monitoring of uplift pressures during remedial works

37 years later from the built of Siurana dam, its drainage system loses their efficiency; as a result, it was necessary to rehabilitate it.

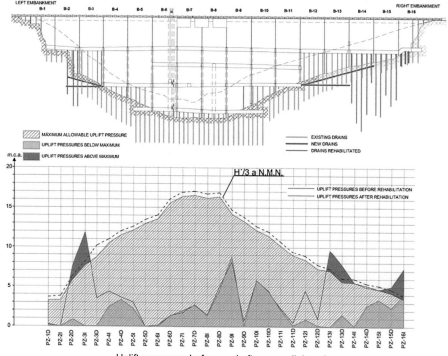

Uplift pressures before and after remedial works

For the rehabilitation works a new methodology was applied to optimize results and reduce costs. The different phases of remedial works were: Geometric analysis, analysis of existing uplift pressures data, installation of a complete piezometer network, analysis of uplift pressures measured, selection of areas to be treated and rehabilitation of drainage network.

LESSONS LEARNT

In Spain there are over 800 gravity dams and a big number of them lack of a reliable system to monitor and control uplift pressures; real sliding safety factor cannot be calculated. In this paper we want to emphasize the benefits, both in safety conditions and reducing maintenance costs, of installing a well-designed system of piezometers. We also include design criteria, backed by experience gained at more than 40 dams (Siurana, La Barca, Tanes, etc.) and a specific methodology, for ease of interpretation and analysis of the data obtained, that has been tested with great results.

Monitoring uplift pressures in dams is a necessity for safety reasons but, as it has been proved in Siurana dam, it is also a very useful tool to save money in the rehabilitation of drainage networks and to check the effectiveness of the rehabilitation as it has been carried out.

DAM MONITORING TO ASSESS VAL DAM BEHAVIOUR DURING THE FIRST YEARS OF OPERATION

Dam type: Roller Compacted Concrete straight gravity dam– 90 m high
Case history category: b, c, f
Main objective: Early detection of failure mechanisms
Main benefit: Correct and on time remedial action
Observations: Horizontal displacements, settlement, seepage, uplift, thermal state

In regard to horizontal displacements, Figure shows the time history for a survey point located on the crest of Val dam, at central block 4. In this case, the deformation of the structure is attributed to the movement of the foundation itself, with very little link to the mechanical loading or deformations of dam body, alignment measurements gave out more than 30 mm after first filling of Val dam, in the downstream direction.

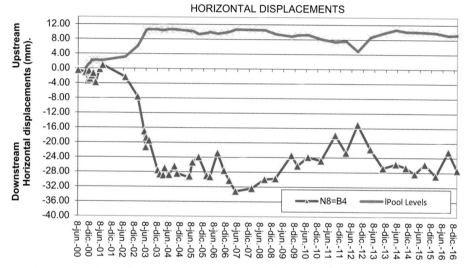

Time history of horizontal displacements at Block 4 of Val dam, from 2000 to 2016 during first filling to present, with reservoir levels included.

LESSONS LEARNT

The main conclusions from the safety inspections at Val dam during the first 15 years of operation are as follows:

1. In general, the dam behaved well during its first filling and the first 15 years of operation, not showing any critical pathology.

2. Maximum vertical settlements reached 7 mm after 15 years of readings, whereas horizontal displacements went beyond 30 mm after first filling, setting a record for all CHE dams.

3. Seepage and drainage rates at Val dam were, and still are, quite low, with a maximum of 2.5 l/s at normal pool levels.

4. Uplift pressures at Val dam also had moderate values, in general, during those first years, showing a distribution with a linear tendency upstream of the drain curtain, and a parabolic one downstream.

IMPLEMENTATION OF A DAM MONITORING MANAGEMENT PROGRAM

Dam type: 55 large dams: arch dams, gravity dams and embankment dams
Case history category: a, b, f
Main objective: Centralization of Monitoring Data
Main benefit: Optimization of Monitoring Data Management
Observations: Integration of Monitoring in the Dam Safety Management Program

The experience gained during the implementation stage and 6 years of operation of a web-based dam monitoring software for the control of more than 55 dams in the Ebro River Basin Authority is shown. At present, a real time control of all automated sensors is done and compared with different threshold values that are established in the operation manual and emergency action plan of each dam. All graphs for periodic and extraordinary safety reports have been individually configured and allow an optimized preparation of safety reviews.

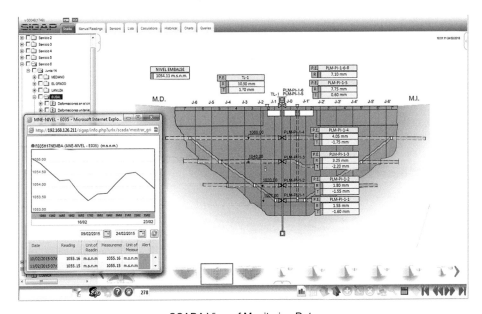

SCADA View of Monitoring Data

The whole life cycle of monitoring data is covered, including data acquisition of manual and automated data, an updated and accessible database, tools for data evaluation and the integration in the owner's dam safety management program.

LESSONS LEARNT

The implementation of a new Data Management System is not just software but the implementation of dam monitoring knowledge which should be carried out by experienced experts. It involves the configuration of the installed monitoring systems, review and upload of formulas and historic data and the preparation of charts that facilitate a target-oriented interpretation of dam behaviour. This usually implies field work and a revision of the installed monitoring system.

A key aspect is the training of the different users, such as engineers, technicians, guards and external consultants.

STORFINNFORSEN POWER STATION - EMBANKMENT DAM IMPROVEMENT INITIATED BY JUST A SMALL CHANGE IN MEASUREMENTS DATA

Dam type: Embankment dam with glacial till core and timber sheet pile - 23m high
Case history category: b, c, d, e
Main objective: Early detection of potential failure mechanisms
Main benefit: Correct and on time remedial action
Observations: Early detection of potential timber deterioration and assessment of remedial works

Total leakage has been small with minor variations (1 to 1.5 l/s) over the years. The readings from the four standpipes have generally indicated constant pressure until 2007, when a slight increase (some 9 mm/year) became evident. Based on these results, in connection to potential deterioration of timber sheet pile and the material in zone 7 (Figure), further investigations started.

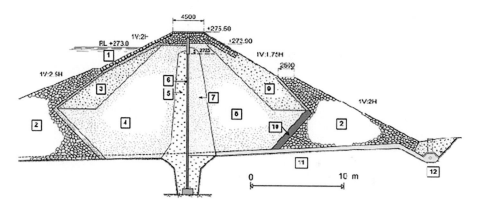

① Rip rap
② Blasted rock
③ Gravel material
④ Support fill, glacial till
⑤ Impervious till core
⑥ Wooden sheet pile
⑦ Transition fill, glacial till
⑧ Support fill, glacial till
⑨ Gravel material
⑩ Transition filters
⑪ Horizontal filter of sand
⑫ Drainage along the toe

Construction section

LESSONS LEARNT

This case shows that also small changes may be important, and *evaluation of data must be set in the context of the construction of the dam*. The small measured change in pressure, together with a not so god original design, was in this case enough to start further investigations. They showed high pore pressure in support fill and that the pressure in the filter layer below the downstream support fill has increased slightly over the last five years or so, indicating possible deterioration in the sealing function of the timber sheet piling.

The deterioration of the timber sheet piles in Storfinnforsen could not be fully determined, although several indications were revealed based on the measurements. Experience from similar dams has shown that the deteriorating process of timber sheet piles has been complete or partly rotten after some decades. *This potential weakness was taken care of by installing a toe berm and additional drainage in the downstream part. The monitoring system was also upgraded* with more sensors for automatic measurements to better understand and detect potential future deterioration of the sealing zone.

IDENTIFICATION OF NEED OF MONITORING AND SURVEILLANCE – A CASE STUDY ON 20 SWEDISH DAMS

Dam type: group of several types of dams
Case history category: d
Main objective: Early detection of potential failure mechanisms
Main benefit: PFMA, definition of monitoring and surveillance needs
Observations: Efficient and PFM-orientated monitoring systems

Fortum is one the largest dam owners in Sweden with about 20 high consequence dams. In order to improve and optimize their monitoring and surveillance activities a project, called "Dam Performance Monitoring", was initiated 2011. The project is focused on the individual need of monitoring and surveillance for each dam (including the gates), monitoring program, staff training/ competence, and organization.

The work started with a questionnaire (based on the questionnaire presented in the Bulletin 158 (DAM SURVEILLANCE GUIDE/GUIDE DE LA SURVEILLANCE DES BARRAGES) in order to define the most important gaps regarding the surveillance work status. Based on the eight main elements in the questionnaire, several sub questions were formulated and used in interviews within the organisation.

Based on the result from the questionnaire it was decided, in a first step, to concentrate efforts on elements 1–5 with the ambition to increase the level with about 25–30 points by the following work:

- PFMA workshop – Identify and estimate the failure modes for the 20 high consequence dams

- Define the monitoring need for those dams and suggest improvements

- Define the monitoring status, including calibration, checking of monitoring devices

- Improve Monitoring Program

- Staff training

- Data collection

The initial work included PFMA for each of the 20 high consequence dams. Workshops were carried out for 16 dams in 2012 and 4 in 2013. Based on the PFMA results the monitoring and surveillance need was summarized in a report. Some additional studies were often initiated in order to achieve more information. The monitoring devices were also tested. With all this work done a new monitoring and surveillance program was set up for the dam. Training of staff has also been done. A WEB-based version of the questionnaire was developed and used after about 75% fulfilment of the first step. The result was promising, showing about 20 points more than the first result.

LESSONS LEARNT

The experience from this work has shown that PFMA is a good basic method with spin offs in several areas such as: Defining monitoring and surveillance need, act as Guidance for operation and maintenance, as well as be adopted for Risk management, and not less important - Inspire and train the organization.

The questionnaire was found useful in order to identify areas for improvement. Actions can be defined and evaluated, and improvements can be verified. This will also encourage the entire organization.

ZIATINE DAM - LEAKAGE THROUGH DAM LEFT BANK

Dam type: Embankment dam with clay core – 32.5 m high
Case history category: e, c, d
Main objective: importance of sealing mats
Main benefit: Temporary remedial action
Observations: Early detection of leakage and assessment of remedial works

The appearance of continuous increase of infiltrations that leads to permanent leakage flow on left abutment was observed. To avoid infiltration bypassing the dam and possible development of internal erosion (suffusion), sealing measures would have been necessary in the left abutment. Unfortunately, the absence of sealing mat has led to design temporary remedial action: a draining trench to ensure efficient collection of infiltration flows as a toe drain design.

Thus, the objective of temporary remedial action was to control the leakage through the left bank, generated due to the presence of sandy horizons with high permeability values **(Lugeon tests = 48 - 80 UL).** Nevertheless, sealing treatment were planned for the semi-permeable alluvial foundation but neglected in the left abutment. Consequently, the fulfilling of the dam is limited not by the normal operating level anymore, but at 7 meters below (around **30m NGT**).

LESSONS LEARNT

The design for embankment dam situated in heterogeneous foundation material should have effective and well monitored sealing treatment at the foundation contact. Drainage body is certainly essential but definitely insufficient without correct sealing that avoid reaching the critical hydraulic gradient, causing irreversible phenomena of internal erosion.

Pro-active surveillance of dams during their life cycle became the standard. The problem was detected from the first dam watering and remedial works were followed with the design offices.

LOW COST, HIGH BENEFIT SHORT-TERM RISK REDUCTION MEASURES IMPLEMENTED BY EFFECTIVE MONITORING AT DORRIS DAM

Dam type: Embankment, 8 m high and 3 km long
Description: Homogenous earthfill
Case history category: a and e
Main objective: Early detection of failure modes
Main benefit: Reduction of risk
Observations: Installation of additional failure mode-oriented monitoring system and adjustment of operating procedures.

Implementation of monitoring changes at Dorris Dam allows for effective routine dam safety surveillance regarding a relatively high-risk potential failure mode until planned dam modification work can be carried out. This case history illustrates that a well-designed monitoring program can achieve substantial risk reduction benefits at very reasonable costs.

Discharge Pipe   **Inlet Structure**

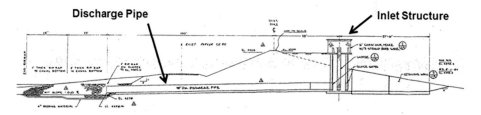

Section view of Dorris Dam spillway/outlet works (along outlet works pipe)

The 2013 evaluation included a recommendation to *"adjust dam operating procedures so that outlet works discharges are halted at least one day each month, so that monitoring for possible seepage flows at or near the spillway and outlet works conduit outfalls can be effectively performed at that time."* Operationally, this recommendation could be implemented without difficulty. Delivery of irrigation water could be delayed one or two days each month without any significant adverse impacts.

Pipe outfalls of Dorris Dam spillway/outlet works during irrigation season

LESSONS LEARNT

Effective potential failure mode analysis work can allow recognition of situations that may present high dam safety risks. Risk analysis work can allow for an improved understanding of those risks. With the understanding gained, it may be possible to implement relatively low-cost monitoring actions and activities that provide substantial dam safety risk reduction, as occurred at Dorris Dam regarding PFM A.4.

PIEZOMETER AT OCHOCO DAM PROVIDES WARNING OF UNSATISFACTORY DAM PERFORMANCE

Dam type: Embankment – 46 m high
Description; Homogenous earthfill
Case history category: e
Main objective: Early detection of failure modes
Main benefit: Detection of seepage related dam failure
Observations: Combination of visual inspection and instrumented monitoring

Piezometer data collected at Ochoco Dam provided the first indication of possible seepage-related dam failure at or near the dam's right abutment. Ultimately a major dam modification project was undertaken to address the identified anomalous performance. Measured water levels in piezometers became much higher than historical performance, leading to a lower dissipation of hydraulic head. In addition, cloudy water in a pond downstream of the dam was observed.

This incident highlights the utmost importance of effective routine dam safety monitoring activities, including both visual surveillance and instrumented monitoring.

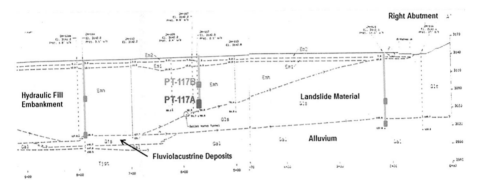

Piezometers installed from the dam crest at the right side of Ochoco Dam

LESSONS LEARNT

The 1989 performance data changed the priority of Ochoco Dam in Reclamation's dam safety program. A restriction on water levels in the reservoir was immediately imposed. In 1991, more right abutment blanketing work was performed, as a temporary, stop-gap action. In 1994, construction work for comprehensive dam modification began. A new core for the dam was constructed at the upstream slope of the dam that had an appropriately designed filter zone downstream of it. The new core for the dam tied into a comprehensive blanket provided for the right abutment of the dam.

Prior to the 1989 anomalous data, Reclamation had in place a plan for future dam safety activities and actions regarding Ochoco Dam. However, instrumentation data collected in 1989, primarily from one piezometer (PT-117A), completely altered those plans. The data indicated more urgent dam safety concerns than had previously been recognized. Without the information from the piezometer, and the actions that resulted, it is not hard to envision that the unrecognized situation at the dam could have worsened, leading to a "break out" of a concentrated seepage flow in the downstream right abutment area, that conceivably could have ultimately led to failure of the dam by breaching, if heroic intervention efforts were not able to arrest the progression of the failure mode.

STEINAKER DAM RIGHT ABUTMENT SLIDE INCIDENT: RESPONSE, EVALUATION, AND SURVEILLANCE

Dam type: Embankment – 50 m high
Description: Zoned earthfill with central clay core
Case history category: e
Main objective: Early detection of failure modes
Main benefit: Detection of landslide and seepage related failure mode
Observations: Installation of additional failure mode-oriented monitoring system

A slide area in embankment material at the upstream right abutment created a new seepage-related potential failure mode at Steinaker Dam. New piezometers were installed to monitor (along with visual surveillance) for possible failure mode initiation/development. New monuments were installed in the slide area to allow effective monitoring for possible new slope movements when the reservoir is drawn down in the future. Aerial surveying was promptly performed to help map the slide area and scarps. A one-day risk evaluation was conducted to assess the risks associated with a new scarp-related seepage failure mode.

Upstream slope at right abutment (slide scarps highlighted in red)

Water levels in observation well close to the landslide (OW-86-6A) did not closely track the reservoir level in months before the slide occurred (as this instrument historically had), but instead the water levels stayed anomalously high. Heavy precipitation and significant reservoir drawdown one month prior to slide detection (which are conditions that could trigger an upstream slide) was observed. When the slide was actually detected, the reservoir was rising and not as much rainfall had occurred in the preceding couple weeks.

LESSONS LEARNT

The nine new sets of slope monitoring monuments allow effective monitoring for new movements in the slide area when the reservoir is being drawn down. Visual monitoring and monitoring of the seven new piezometers will allow effective monitoring for the new seepage-related failure mode associated with the right abutment slide area. These instruments will allow effective monitoring until repairs to the slope can be made.

WANAPUM DAM SPILLWAY INCIDENT: RESPONSE, ROOT CAUSE, REMEDIATION, AND SURVEILLANCE

Dam type: Gravity and embankment flanks – 56 m high
Description: Concrete and zoned earthfill
Case history category: e
Main objective: Early detection of failure modes
Main benefit: Detection of fracture in upstream side of spillway
Observations: Combination of visual inspection and instrumented monitoring

Excessive tensile stresses at the upstream face of an ogee spillway structure caused a long, continuous, horizontal crack that allowed uplift pressures within the crack to result in significant displacements of one section of the structure, that were detected by visual surveillance efforts. Surveys indicated that the top of Pier 4 had moved downstream about 5 cm. Ultimately a major dam modification project was undertaken to address the situation, which included enhanced surveillance and monitoring efforts during the various portions of the construction process.

Instrumentation was added during spillway construction to complement and enhance existing instrumentation, surveillance, and monitoring procedures at the Dam. Instrumentation installed during spillway remediation included lift joint drains in each monolith (over 80 lift joint drains drilled through the ceiling of the spillway gallery), vibrating- wire piezometers installed through the floor of the gallery, vibrating-wire crack gages installed at a horizontal lift joint in Monolith 4, and 2-D crack gages installed at contraction joints in the gallery.

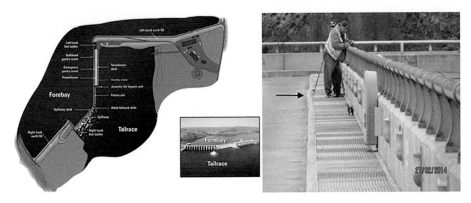

Plan of Wanapum Dam and offset at upstream curb of Pier 4

LESSONS LEARNT

The Wanapum Spillway fracture was a significant dam safety incident that required the efforts of numerous personnel internal to GCPUD, from outside consultants, and from the regulator (the FERC) in order to provide an adequate and appropriate response to recover from the incident in a timely fashion. The early establishment of recovery goals and identifying the resources necessary to accomplish these goals allowed GCPUD to recover from this incident in a timely fashion.

WOLF CREEK DAM: USING INSTRUMENTATION DATA TO ASSESS BARRIER WALL EFFECTIVENESS

Dam type: Gravity and embankment– 79 m high
Description: Concrete and earthfill
Case history category: e
Main objective: Early detection of failure modes
Main benefit: Evaluation of effectiveness of barrier wall
Observations: Statistical analysis of monitoring data

Built in the 1940's, the clay embankment of Wolf Creek Dam was placed directly on an untreated karst foundation. The dam has undergone past remediation, including emergency grouting in the 1960's and the installation of a barrier wall through the crest of the dam and near the switchyard in the 1970's. Information gathered since the 1970's indicated that the 1970's barrier wall was not deep enough or long enough to adequately reduce risks. Recently, distress indicators observed on site include increased piezometer levels and the construction of a 1.16-km-long, 84m deep, secant pile barrier wall designed to intercept large karst features in the dam's foundation. Afterwards, the dam was subjected to a Post-Implementation Evaluation to reassess the risk level of the project, using a multivariable statistical analysis of the piezometer data. Correlation plots of the pool vs. piezometer response were used to identify piezometers that demonstrated a high statistical significance to headwater. Piezometers that were identified as having statistical significance all had R^2 values of greater than 0.35.

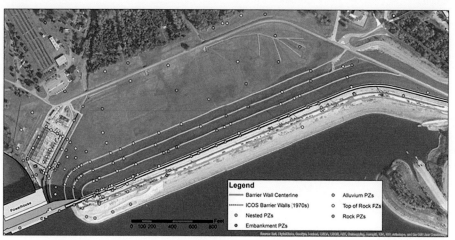

Plan view showing the distribution of the 239 piezometers installed at the dam

LESSONS LEARNT

Despite the lack of baseline data and transient construction responses, the multiple linear regression analysis indicates that the completion of the dam safety modification project at Wolf Creek Dam successfully reduced the influence of headwater on downstream piezometers. In some cases, headwater was no longer a statistically significant predictor in post-construction models. In other cases, the sign on the regression coefficient went from positive to negative indicating an inverse relationship between headwater and piezometer response. This could be an indication of additional lag in the system resulting from a lengthened seepage path. The validity of these findings will be evaluated long-term as data is collected during multiple cycles of normal (non-restricted) pool operation.

7. KEYWORD INDEX

KEY WORDS	CASE HISTORIES
Abutment	Santa Rita, Urrá I
Advection	Canadian embankment dam with central till core (case history n°10)
Ageing	La Paliere
Alkali-aggregate reaction (AAR)	Cahora Bassa, Song Loulou
Amortization period	La Minilla + El Gergal
Analysis	Automatic monitoring system of Portuguese dams (case history n°45)
Analysis of alternatives	La Minilla + El Gergal
Annual thermal response	Canadian embankment dam with central till core (case history n°10)
Asphaltic concrete core	Storvatn
Atypical leakage event	Kootenay Canal Forebay
Automated monitoring	Automatic monitoring system of Portuguese dams (case history n°45), Okuniikappu, Zillergründl, 55 dams in Ebro river basin
Basin	El-Karm
Bathymetry	Isola Serafini
Behaviour	20 high risk dams
Bio-location	Ohrigstad
Bitumen	Trial dam
Canal	La Paliere
Climate change	Sylvenstein
Collapse	Tuizgui Ramz
Compaction grouting	WAC Bennett (trends analysis)
Concrete	Ambiesta
Concrete joint leakage	Kootenay Canal Forebay
Concrete swelling	Cahora Bassa, Kouga, Song Loulou
Construction	Mirgenbach
Construction control data	Canadian embankment dam with central till core (case history n°10)
Control gallery	Sylvenstein
Control method	El Chocón
Corrosion	Folsom
Crack	Karun 4
Crack pattern	Belfort
Cracking	Masjed-e-Soleiman, Yashio, Zillergründl
Cracks	Ambiesta, El-Karm, Katse, Poiana Uzului, Zeuzier
Culvert joints	Oued El Makhazine (culvert joint)
Cut-off	Oued El Makhazine (cutoff wall)
Cutoff wall	La Loteta
Cut-off wall	Sylvenstein
Dam	Porce II
Dam breach	Möhne, Eder and Dnieprostroi dams
Dam busters	Möhne, Eder and Dnieprostroi dams

(Continued)

KEY WORDS	CASE HISTORIES
Dam core heterogeneities	Canadian embankment dam with central till core (case history nº10)
Dam failure	Belfort, Teton, Tous, Zoeknog
Dam operation	Automatic monitoring system of Portuguese dams (case history nº45), Durlassboden
Dam safety	Caspe II, La Loteta, Val
Data base	Several dams (MISTRAL)
Deformation	Tona, Urrá I
Deformation measurement	Automatic monitoring system of Portuguese dams (case history nº45), Val
Deformations	La Aceña, Masjed-e-Soleiman, Trial dam, Zeuzier
Design calculation error	Wanapum
DGPS system	La Aceña
Differential settlement	El-Karm, Kootenay Canal Forebay
Displacement	Okuniikappu
Disposal	Urrá I
Divers	Ohrigstad
Drain materials	Oued El Makhazine (cutoff wall)
Drainage	Gura Raului, La Loteta, Poiana Uzului, Santa Rita, Ziatine
Drainage drills	Paltinu
Drainage gallery	Muravatn
Drainage network	Siurana
Drainage system	Storfinnforsen
Drains	Siurana
Drilling	Driekoppies
Earthfill dam	Gotvand
Earthquake	Yashio
Effective visual monitoring	Dorris Dam
Electromagnetic survey	Canadian embankment dam, 1200 m long (case history nº11)
Embankment dam	La Loteta, Ziatine
Emergency plan	La Loteta
Energy balance orifice flow equation	Kootenay Canal Forebay
Erosion	Isola Serafini, La Loteta, Teton, Tuizgui Ramz, Urrá I
Excavation	Cortes
Extensometer	Gmuend
Failure	Malpasset, Porce II, Porce II, Vajont
Fault zone	Muravatn
Fill	Urrá I
Fine content	Canadian embankment dam with central till core (case history nº10)
Finite element model	Kouga
Finite elements method	Zillergründl
First impounding	Gotvand
Flood	Tous

(Continued)

KEY WORDS	CASE HISTORIES
Flow path	Ohrigstad
Foundation	Belfort, Malpasset, Mšeno, Val, Zeuzier
Foundation clay layer	Canadian embankment dam, 1200 m long (case history n°11)
Foundation sand layer	Canadian embankment dam, 1200 m long (case history n°11)
Foundation treatment	Caspe II, Durlassboden, La Loteta
Gauge	Mirgenbach
Geneva Convention	Möhne, Eder and Dnieprostroi dams
Geographic Information System (GIS) software	WAC Bennett (ROV)
Geomembrane	Pecineagu
Geomembrane liner	Kootenay Canal Forebay
Geostatistics	Canadian embankment dam with central till core (case history n°10), Canadian embankment dam, 1200 m long (case history n° 11)
Geotechnical engineering	Zelazny Most (Poland)
Geotechnical investigations	Malpasset, Storfinnforsen
Geotextile	Oued El Makhazine (culvert joint)
Glacial	Sylvenstein
Grout curtain	Paltinu
Grout curtain	Durlassboden
Grouting gallery	Mšeno
Heat of hydration	El-Karm
Heightening	La Paliere
High resolution imagery	WAC Bennett (ROV)
Hydraulic conductivity	Canadian embankment dam with central till core (case history n°10)
Hydraulic gradient	Teton, Ziatine
Impervious blanket	La Loteta
Incident detection	Caspe II, La Loteta
Infiltration	Porce II
Inspection	Etang, Porce II
Installation	Inyaka
Instrument performance evaluation	WAC Bennett (trends analysis)
Instrument reading creep	WAC Bennett (trends analysis)
Instrument reading lag time	WAC Bennett (trends analysis)
Instrument reading performance bound	WAC Bennett (trends analysis)
Instrumentation	Katse, Kouga, Svartevann, Tona, Viddalsvatn
Intentional dam demolition	Möhne, Eder and Dnieprostroi dams
Internal erosion	Canadian embankment dam with central till core (case history n°10), Canadian embankment dam, 1200 m long (case history n° 11), Caspe II, El Chocón, La Loteta, Svartevann, Viddalsvatn, Ziatine

(Continued)

KEY WORDS	CASE HISTORIES
Jointmeters	Seymareh
Joints	El-Karm, Zeuzier
Karst	La Loteta
Karst foundation	Wolf Creek
Key driver variable	WAC Bennett (trends analysis)
Landslide	Cortes, Grand Maison Reservoir, Urrá I, Vajont
Large scale test	Tona
Leak detection	Kootenay Canal Forebay
Leakage	Etang, Kyogoku Upper Reservoir, Ohrigstad, Pecineagu, Storfinnforsen, Svartevann, Tuizgui Ramz, Viddalsvatn, Ziatine
Lidar survey	Canadian embankment dam, 1200 m long (case history n° 11)
Liquefaction	Santa Rita
Masonry dam	Mšeno
Mathematical model	Zillergründl
Measurement system	Kyogoku Upper Reservoir
Membrane	Storvatn, Trial dam
Model	Gura Raului
Modulus	Tona
Monitoring	Caspe II, Cortes, Driekoppies, Durlassboden, El Chocón, Etang, Grand Maison Reservoir, Gmuend, Inyaka, Katse, Kouga, La Aceña, La Loteta, La Paliere, Mirgenbach, Muravatn, Pecineagu, Porce II, Porce II, San Giacomo, Santa Rita, Several dams (MISTRAL), Siurana, Svartevann, Sylvenstein, Val, Viddalsvatn, Yashio, Zillergründl, 20 high risk dams
Monitoring system	Cahora Bassa, Poiana Uzului, Tuizgui Ramz, Zeuzier
More thoroughly installation	Alborz
Multibeam sonar	WAC Bennett (ROV)
Numerical model	Zillergründl
Operation	Val
Optical fiber	Storfinnforsen
Overtopping	Tous
Pendulums	Katse, Seymareh
Performance	Porce II, Porce II
Piezometer	Caspe II, Durlassboden, La Loteta, Santa Rita, Urrá I, Zillergründl, Zoeknog
Piezometer measurements	San Giacomo
Piezometer network	Siurana
Piezometers	Ochoco
Piping	Viddalsvatn, Ziatine, Zoeknog
Pneumatic drills	Driekoppies
Pore pressure	Canadian embankment dam, 1200 m long (case history n° 11), Caspe II, Durlassboden, Inyaka, La Loteta, Malpasset, Santa Rita, Svartevann, Ziatine, Zillergründl
Pore pressures rise	Storfinnforsen

(Continued)

KEY WORDS	CASE HISTORIES
Potential failure mode analysis	20 high risk dams
Potential failure modes	Dorris Dam, Ochoco, Steinaker, Wanapum
Pressure	Muravatn, Trial dam
Pressure cells	Driekoppies
PVC waterstop	Kootenay Canal Forebay
Quarry	Cortes
Redundant energy source	Tous
Regression analysis	WAC Bennett (trends analysis)
Rehabilitation	El Chocón, Karun 4, Santa Rita, Siurana
Reinforcement work	Yashio
Relief wells	Canadian embankment dam, 1200 m long (case history n° 11), Oued El Makhazine (cutoff wall)
Remedial work	Siurana, Ziatine
Remotely operated underwater vehicles (ROV)	WAC Bennett (ROV)
Repair	La Loteta
Repair work	Kyogoku Upper Reservoir
Reservoir level correlation	WAC Bennett (trends analysis)
Reservoir slope	Grand Maison Reservoir
Responding to incident	Steinaker
Riprap	Svartevann
Riverbed	Isola Serafini
Rockslide	Katse
Roller compacted concrete	Porce II
Safety	Etang, Grand Maison Reservoir, La Paliere, Mirgenbach, Porce II, Porce II, Santa Rita, Several dams (MISTRAL), Siurana
Safety of dams	Automatic monitoring system of Portuguese dams (case history n°45), 55 dams in Ebro river basin
Saturation	Urrá I
Seepage	Caspe II, Comoé, La Loteta, Mšeno, Poiana Uzului, Teton, Val, Ziatine
Seepage along conduit	Dorris Dam
Seepage control	Paltinu
Seepage detection	Canadian embankment dam, 1200 m long (case history n° 11)
Seepage flow monitoring	WAC Bennett (trends analysis)
Seepage flow-reservoir level correlation	Kootenay Canal Forebay
Seepage paths	Comoé
Seepage through ancient landslide	Ochoco
Seepage through foundation	Wolf Creek
Seepage weir flow	WAC Bennett (trends analysis)

(Continued)

KEY WORDS	CASE HISTORIES
Seismic resistance	Zillergründl
Settlement	La Paliere, Porce II, Svartevann, Val, Zeuzier
Side scan sonar	WAC Bennett (ROV)
Sinkhole	Viddalsvatn
Sinkhole remediation	WAC Bennett (trends analysis)
Slab-plinth joint	Kootenay Canal Forebay
Sliding in body of concrete dam	Wanapum
Slope stability	Mirgenbach, Vajont
Sonic tomography	Ambiesta
Spillway gate failure	Folsom
Spillway gates	Tous
Stabilizing berm	Storfinnforsen
Standpipe destruction	Alborz
Statistical method	55 dams in Ebro river basin
Strain	Storvatn
Stress	Storvatn
Sulphate attack	Song Loulou
Supplementary instrumentation sections	Gotvand
Surveillance	Gura Raului, Siurana, Svartevann, Viddalsvatn
Tailing dam	Zelazny Most (Poland)
Tainter gate	Folsom
Temperature	Canadian embankment dam with central till core (case history n°10), Gmuend, Gura Raului, Poiana Uzului, Storfinnforsen, Val
Thermal loading	Seymareh
Threshold	Several dams (MISTRAL)
Thresholds	Porce II
Till core	Canadian embankment dam with central till core (case history n°10)
Timber sheet pile	Storfinnforsen
Total pressure cells	Inyaka
Training	20 high risk dams
Transient conditions	Masjed-e-Soleiman
Trend analysis	WAC Bennett (trends analysis)
Trunnion coefficient of friction	Folsom
Turbid leakage	Oued El Makhazine (culvert joint)
Underground seepage	Comoé
Underwater inspection	WAC Bennett (ROV)
Underwater survey	WAC Bennett (ROV)
Upgrade monitoring	La Minilla + El Gergal
Uplift	Porce II, Val
Uplift pressure	Canadian embankment dam, 1200 m long (case history n° 11), Siurana
Upstream blanket	La Loteta

(*Continued*)

KEY WORDS	CASE HISTORIES
Upstream facing	Etang
Upstream method	Zelazny Most (Poland)
Upstream slope failure	Steinaker
WAC Bennett dam	WAC Bennett (trends analysis)
Water pressure data	Wolf Creek
Weathered rock	Comoé
Weir flow	Kootenay Canal Forebay